People and Biodiversity Policies

IMPACTS, ISSUES AND STRATEGIES FOR POLICY ACTION

by

Philip Bagnoli, Timo Goeschl and Eszter Kovács

ORGANISATION FOR ECONOMIC CO-OPERATION AND DEVELOPMENT

The OECD is a unique forum where the governments of 30 democracies work together to address the economic, social and environmental challenges of globalisation. The OECD is also at the forefront of efforts to understand and to help governments respond to new developments and concerns, such as corporate governance, the information economy and the challenges of an ageing population. The Organisation provides a setting where governments can compare policy experiences, seek answers to common problems, identify good practice and work to co-ordinate domestic and international policies.

The OECD member countries are: Australia, Austria, Belgium, Canada, the Czech Republic, Denmark, Finland, France, Germany, Greece, Hungary, Iceland, Ireland, Italy, Japan, Korea, Luxembourg, Mexico, the Netherlands, New Zealand, Norway, Poland, Portugal, the Slovak Republic, Spain, Sweden, Switzerland, Turkey, the United Kingdom and the United States. The Commission of the European Communities takes part in the work of the OECD.

OECD Publishing disseminates widely the results of the Organisation's statistics gathering and research on economic, social and environmental issues, as well as the conventions, guidelines and standards agreed by its members.

Also available in French under the title:
Aspects redistributifs des politiques pour la biodiversité

Corrigenda to OECD publications may be found on line at: *www.oecd.org/publishing/corrigenda.*

Foreword

The need for protecting and safeguarding biological diversity on this planet as the size and impact of human population expands is increasingly well understood. Significant efforts are underway all over the world to save endangered species, protect ecosystem services, and conserve vulnerable genetic diversity.

Implementing policies to make economic development more compatible with the sustainability of biodiversity is a relatively new endeavour that has met with many successes, as well as a few setbacks. The experience with those policies illustrates that the question of equity and the distribution of impacts is not only an important characteristic of policy, rather it is a central issue in the policy's success or failure. The lessons from the early examples of implemention will need to be learnt swiftly as new sets of policies are being designed to manage the pressures from agricultural expansion, population growth, infrastructure development and climate change on biodiversity and essential ecosystem services. These pressures are likely to lead to the need for even more forceful policy to protect biodiversity – and making distributive impacts even sharper. The OECD Environmental Outlook to 2030, *for example, features a baseline economic scenario that sees world GDP doubling by 2030 and the addition of almost two billion people. That increased population represents more than the combined current population of all OECD countries.*

This book undertakes a timely review of economic issues related to distributive impacts of biodiversity policy. It explores the wide range of those impacts and how they happen. It also illustrates a great deal of practical experience in dealing with them so as to provide policy makers with numerous examples of successful strategies.

This book is the culmination of a work programme by the OECD Working Group on Economic Aspects of Biodiversity (WGEAB) to examine distributive issues in biodiversity policy. Of particular interest for the work was the ways in which distributive impacts were impeding successful implementation of biodiversity policies. An important milestone in that work programme was a workshop held in Oaxaca, Mexico, in April 2006 that was sponsored by the government of Mexico. A clear need was identified at the workshop for a wider treatment of distributive issues, including discussion of underlying issues. Accordingly, the scope of this book is to be as comprehensive as possible, without being formal in its exposition.

Under the WGEAB's guidance, this book was drafted with contributions from Timo Goeschl (University of Heidelberg, Germany), Eszter Kovács (Ministry of

Environment and Water, Hungary) and Philip Bagnoli (OECD Secretariat). Fiona Hall provided excellent editorial assistance, and Jane Kynaston provided excellent support to the production of the book.

This book is published under the responsibility of the Secretary-General of the OECD.

PEOPLE AND BIODIVERSITY POLICIES – ISBN 978-92-64-03431-0 – © OECD 2008

Table of Contents

List of boxes

PEOPLE AND BIODIVERSITY POLICIES – ISBN 978-92-64-03431-0 – © OECD 2008

List of tables

List of figures

Executive Summary

What are distributive effects and why do they matter?

Policies to maintain and improve biologically diverse habitats and ecosystems aim to benefit society as a whole by realising all of biodiversity's values: material benefits, as well as those that are less easily quantified. However, like any other environmental policy, while biodiversity policies can improve aggregate well-being, they can also create winners and losers. For example, in developed countries limitations on land use to protect biodiversity can sometimes reduce income to the individual landowner, but deliver benefits to the general public. In developing countries where natural resources are an important livelihood source for people, biodiversity protection can reduce access to these natural resources, thus imposing a cost on poorer people. These are the "distributive effects" of biodiversity policies and they are important to understand for the following reasons:

- Ignoring the distributive impacts can imperil a policy that would otherwise be beneficial for the general public.
- Policy-makers are under increased pressure to demonstrate that their policies are informed by and comply with criteria emanating from global policy discourses such as the Millennium Development Goals. These criteria frequently contain explicit distributive objectives with little or no guidance on how to fit biodiversity policies around them.
- OECD policy guidelines (OECD, 1997) require policy-makers to assess the distributive effects of policy interventions on the absolute and relative well-being of different groups of people.

Distributive effects are conceptually different from efficiency effects (gains) in biodiversity or other environmental policies. A policy's efficiency effects are the benefits or welfare gains that a policy achieves over and beyond the costs incurred. Traditionally, separating distributive from efficiency effects has been a hallmark of economic analysis. The reason is simple: efficient policies will maximise the benefits from a given level of inputs of natural resources, capital and labour. These maximised social gains can then be redistributed according to society's preferences. However, this book argues that such separation is often not useful for biodiversity policies (see below).

The aim of this book is to help policy-makers who are responsible for designing and implementing biodiversity policies understand the relevance of distributive issues (for both equity and efficiency), and how to integrate them into policy formulation. In this book we analyse the distributive impacts of biodiversity policies across different groups, geographical scales and time. We describe the methods for measuring distributive effects and explain the relationship between policy objectives, instrument choice and distributive outcomes. We provide arguments for integrating distributive issues into biodiversity policies. We also offer different methods for addressing distributional concerns in policy-making and for managing conflicts induced by biodiversity policies. Finally, we present a wealth of case studies to document both the complex chains leading to distributive outcomes and best practice in merging efficiency and equity considerations in policy design, implementation, and management.

The distribution of costs and benefits over time and space

In order to analyse the distributive effects of biodiversity policies, the benefits and costs of the policy first need to be identified. We also need to understand how benefits and costs vary across different groups geographically and over time.

The implementation of biodiversity policies has implications both for international equity between countries and between very different economic groups. This is because any costs of implementing biodiversity policies are frequently concentrated locally – in those areas where biodiversity is actually managed and to those people who can afford them least. For example, costs may be borne by those whose access is restricted to the protected biodiversity or whose property is damaged by an increase in biodiversity. At the same time, many of the policy's benefits may be felt many hundreds or thousands of miles away, by individuals or groups who depend less directly on the protected area.

Local benefits *can* be significant, but this depends on the policy mode applied. For example, management regimes for protected areas which ensure income to local people from tourism can compensate them for livelihood losses. Clearly defined systems for encouraging or allocating financial benefits for locals can underpin that outcome.

The varying time scales over which biodiversity policies are felt also create distributive impacts that need to be studied. Policy decisions today may affect

 PEOPLE AND BIODIVERSITY POLICIES – ISBN 978-92-64-03431-0 –

individuals currently alive differently to future generations. The policy-making process therefore needs to compare benefits and costs of biodiversity policies that may arise at vastly different points in time and justify them against some measure of intergenerational equity. But comparisons across time raise questions about how to measure trade-offs both within a generation, as well as across generations. For example, if the costs have to be incurred right away (*e.g.* curtailing economic growth now) while the benefits occur at some stage in the future (reduced global warming), then how do we compare flows at such different points in time?

Who benefits from and who pays for biodiversity policies?

Biodiversity and ecosystems are important as key productive assets (*e.g.* fish, or timber), for the "services" they provide (*e.g.* as carbon sinks or in water purification) now or in the future, and for the sheer pleasure we get from their continued existence (aesthetic values, cultural values, etc.). Economists use distinct value categories to capture these various sources of biodiversity's contribution to human wellbeing, with the most fundamental categories being those of use and non-use values. Taken together, these value categories make up the total economic value (TEV) of biodiversity, *i.e.* the total contribution of biodiversity to humanity (Pearce and Moran, 1994). The concept of the TEV allows us to evaluate the benefits of policies that affect the availability of biodiversity. It does so by assessing the changes in the values within each value category of the TEV that occur as a result of the policy. When a policy sacrifices more benefits of biodiversity than are gained from its loss at the margin, then this policy should not be allowed to proceed.*

Against the list of benefits identified through TEV, we must compare the costs – monetary and otherwise – of maintaining/procuring these goods and services through biodiversity protection. For policy-makers to decide which policy is the most appropriate, these costs of biodiversity policies also need to be accounted for. The costs of biodiversity policies can be categorised into:

- **Direct costs** of implementing the policy, *e.g.* budgetary expenses raised through taxation. These costs tend to affect governments and are generally smaller than other costs.
- **Indirect costs:** *e.g.* crop losses at the boundaries of protected areas as a result of increased wildlife population levels. Exposure to these costs will be

* That is, the incremental social loss (material and non-material) should be offset by the incremental gain from the reduction of biodiversity.

higher for those more reliant on extractive and consumptive activities in, or adjacent to, a conservation area.

- **Opportunity costs:** the value of lost consumption possibilities previously exercised and no longer possible, or of future consumption possibilities. These opportunity costs are the main costs associated with biodiversity policies.

We also need to remember that households with different incomes rely on very different goods and services generated by biologically diverse ecosystems and habitats. The most important interaction between low income households and their natural environment in developing countries is through extractive and consumptive activities. Richer households are more likely to be interested in the public goods aspects of biodiversity (aesthetic values, ecosystem services, etc.) as their income is less likely to depend directly on primary resources.

Though the results are mixed, research generally shows that biodiversity policies that enhance the supply of biodiversity-related goods and services will typically generate greater benefits for the better-off, and sometimes impose net costs on the less well-off. Furthermore, biodiversity is mostly (though not exclusively) found in developing countries, where income levels are somewhat lower than in most OECD countries. In many cases a significant share of the non-use benefits of conserving biodiversity in these countries might accrue to developed countries. This asymmetric distribution is another key dimension of distributive issues at the international level.

Policy type and mode also have distributive effects (Table 0.1):

- **Voluntary *versus* non-voluntary policies.** Voluntary policies, such as conservation easements or payments for ecosystem services, allow potential participants to decide whether to contribute to the policy or not. Non-voluntary biodiversity policies force individuals to participate in the policy. Examples are restrictions on property rights, *e.g.* by banning land development; or taxes and fees, *e.g.* a pesticides tax. On the one hand, non-voluntary instruments will – as a rule – produce individual losers, creating incentives for losers to undermine the policy. On the other hand, such policies can generate significant net benefits at the aggregate level (despite the few individual losers). Handling this trade-off between the creation of losers and viability of the policy in the face of losers' opposition is one of the key challenges in the design of conservation policies that this book explores.
- **Reward-based *versus* property-based policies.** Reward-based policies leave it to the policy participant to decide how much of a certain activity is carried out, but specify a fee that typically increases with the volume of the activity being carried out. Property-based policies leave it to the market to

Table 0.1. **Classification of policy instruments**

Policy mode	Participation	
	Voluntary	Involuntary
Change in property and use rights	***Type II*** Land purchase Conservation easements	***Type IV*** Designation of protected areas Land use regulations Trade restrictions
Distributive effects	*No evidence of losers; however, some gain more than others, i.e. those with more assets, especially land*	*Sharp reduction of access or livelihood for some; enhancement of livelihood for others* *Gain in indirect benefits for larger number of people*
Change in rewards	***Type I*** (Public) payments for ecosystem services Market creation Product certification	***Type III*** Biodiversity-related taxes User fees Removal of perverse incentives
Distributive effects	*Few people suffer direct losses from the policy, but some will see relative prices change in the market, which may affect them adversely*	*There will be losers if the welfare gains from an increase in biodiversity are less than the increase in taxes an individual must pay to finance the policy*

determine the value of the activity, but ensure that the conditions for the market to do so are present.

The relationship between instrument choice and distributive impacts shows that policy-makers have considerable scope for assigning the benefits and costs to different groups depending on which instruments they choose. However, there are trade-offs between the desirability of being able to fully implement policy (which calls for coercive instruments) and the desirability of being able to avoid creating a high volume or a high individual incidence of policy losers (which calls for voluntary instruments). Historically, this trade-off has led to a strong bias towards policies that combine coercion with changes in property rights. The result has often been problematic distributive outcomes and a failure of the policy on the ground. Other approaches, such as tax-based measures, seem underexploited in their potential to strike a middle ground in this trade-off.

How do we measure the distributive impacts of biodiversity policies?

The Table 0.2 compares the most important methods for measuring the distributive effects of biodiversity policies, and describes when they are most appropriate.

Deciding which method to choose depends on the policy measure, the geographical scale, and data availability. Each of the methods has particular strengths in terms of capturing distributive effects, and weaknesses in terms of either omitting important dimensions or not allowing some type of aggregation to be carried out.

How do we avoid the distributional impacts of biodiversity policies?

According to the basic model of welfare economics, policies aimed at correcting externalities (such as biodiversity policies) should be separate from policies with redistributive objectives. Separating equity and efficiency objectives leaves biodiversity policies unencumbered by additional constraints and obligations and free to pursue those policy options that promise to deliver the greatest social gains. These maximised social gains generated by the biodiversity policy are then available for redistribution to those made worse-off by the policy. However, for biodiversity policies there are a number of fundamental and practical reasons why such separation is not always possible, and why implementing biodiversity policies which do not incorporate distributive aspects may involve serious efficiency losses:

- The "public good" nature of many of biodiversity's goods and services.
- The transaction costs of moving a dollar from one person to another.
- Incomplete information about the nature of the policy and its impacts.
- Frequent lack of geographical overlap between the winners and losers and the scope of the institution making the transfers.
- The nature of common property resource management systems, where a small redistribution from rich to poor can induce a collapse of conservation efforts if the need for policy has not been made clear and hostility to such efforts is induced.
- The changes in distributive impacts of biodiversity policies over time: there are few functional mechanisms for transfers across generations.

 PEOPLE AND BIODIVERSITY POLICIES – ISBN 978-92-64-03431-0 – © OECD 2008

Table 0.2. **Advantages and disadvantages of the key methods for measuring distributive effects of biodiversity policies**

Method	Strengths	Weaknesses	Applications	Examples
Measures of equality of income distribution (Lorenz curve, Gini coefficient)	Graphical illustration and numerical measure of equality	Cannot be used in a very complex situation. Large statistical data requirements	Can be used for a well-defined group	Measures of equality of income distribution impacts of privatisation of mangroves in Viet Nam
Extended cost-benefit analysis (by stakeholder groups)	Gives quantitative results segregated by stakeholder groups	Extensive statistical data are needed	At both national and local level, where income or stakeholder groups can be easily identified. Where monetary valuation of both costs and benefits is possible	Three potential park scenarios in the Ream National Park, Cambodia. Three scenarios in the Leuser National Park, Sumatra
Social accounting matrix (SAM)	Shows the flow of income from one sector to another	Extensive statistical data are needed; rather complicated method. Problematic where no financial information is available	Can be used in local, regional and national circumstances	Distributional impacts of alternative forest management of the Upper Great Lakes region, USA
Distributional weights	Can compare efficiency and distributive impacts on a common scale	Extensive statistical data and assumptions about utility function are needed	Both national and local level where income groups or stakeholder groups can be easily identified. Where monetary valuation of both costs and benefits are possible	*UK Green Book*
Atkinson inequality index	Uses normative judgements about social welfare	Has been used only in the narrow field of income studies. Applicability to biodiversity policies is still an open question	Can be used at international, national, and local levels to the extent that normative judgements can be plausibly applied in the chosen context	No example related to biodiversity policy yet
Employment-based analysis	Unconventional, straightforward measure of the level of employment	Income changes cannot be measured by this method. Cannot capture other social effects. May hide important dimensions of job status, qualification match and labour market frictions	In rural areas, where employment changes are more important than the changes in the income level	Measuring the employment impact of different farming activities, Yucatan, Mexico

Table 0.2. **Advantages and disadvantages of the key methods for measuring distributive effects of biodiversity policies** *(cont.)*

Method	Strengths	Weaknesses	Applications	Examples
Child nutritional health status	Unconventional, straightforward measure of the nutritional status of children	Nutritional status depends on many factors. Former calculations had statistical and other problems	In developing countries, where it is difficult to use other measures and where nutrition can have direct link to biodiversity policy measures	Measuring nutritional status of children in marine protected areas in the Philippines
Stochastic dominance analysis	Multidimensional analysis of the distribution of social welfare (not only income and wealth, but others, *e.g.* education or health)	Strong assumptions are needed about how the dimensions relate to welfare, and social weights need to be given to different dimensions	At local, regional or national level	The Human Development Index of the United Nations
Multi-criteria analysis (MCA)	A wide range of distributive effects can be measured using social and economic criteria: the level of measurement is not a problem. It can be a base for further discussion with stakeholders and for assessing trade-offs	Results heavily depend on the weights given to the criteria (weights can be given by experts, stakeholders or policy-makers)	In local, regional and national situations, and in complex cases when many criteria need to be taken into account, and where some of the effects cannot be measured in monetary terms	Australian forest policy. Buccoo Reef Marine Park, Tobago. Šúr wetland nature reserve, Slovakia
Social impact assessment (SIA)	Impacts on stakeholders and distributive effects are assessed. All other methods can be used for the assessment	Sometimes it is superficial and lacks monetary data (if there is no clear guidance or indicators given)	In any policy situations at local, regional and national level	Stakeholder analysis in the Royal Bardia National Park, Nepal.
Hyperbolic discounting techniques	Allow future generations to be explicitly considered. Potentially reduce distributive impacts across generations	Can be inconsistent since a decision taken today would not be taken tomorrow, even if nothing changes	Any policy whose effects will last more than about 10 years	*UK Green Book*

- Political economy considerations: entrenched interests and political power cannot easily be separated from the policy itself and can have a range of distributive impacts.
- Conflict, which can be costly and expensive to resolve.

Some of these efficiency losses will be very palpable, as in the case of conflicts arising around conservation policies. Others will be large, but only evident to future generations. Some may be difficult to foresee at the time of planning, since groups, individuals and institutions will change their behaviour over

 PEOPLE AND BIODIVERSITY POLICIES – ISBN 978-92-64-03431-0 – © OECD 2008

time. But they all imply that biodiversity policy approaches which consider distributive issues from the start are likely to be more efficient and effective.

Integrating efficiency and equity into biodiversity policies

This study suggests four approaches for integrating distributive issues into biodiversity policies, in increasing order of complexity and hand-over of control by the policy-maker:

- Methodological: use the methodologies identified in this book (Chapters 2 to 4) to better understand and account for the welfare impacts in policy design. This means that the policy-making process is now augmented by a consideration of distributive impacts. At the same time, the policy-maker still retains full control over information gathering, policy evaluation and choice, and instrument choice.
- Procedural: enrich the policy-making process by using consultative and participatory approaches to involve those who will be directly affected by biodiversity policies. Effective consultation allows various groups to express their views so potential conflicts can be addressed and acceptable solutions developed.
- Institutional: accompany biodiversity policies with explicit changes to the institutional structure under which individuals and groups take decisions that affect the target habitats and ecosystems. These may include creating property rights and entitlements as well as novel markets and contract schemes in order to manage distributive impacts.
- Combined: bring together the second and third approaches above. Thus, institutional changes allow affected individuals, households and groups to become actively involved in policy decision-making on an ongoing or even permanent basis. In its most extensive form, this includes measures that devolve to individuals or groups affected part of the management of the policy. Using participatory methods in the design of the biodiversity policy measure can help identify and mitigate the distributive effects of the policy measure in a way that is satisfactory for the public and for the stakeholders. It can also prevent conflicts and secure the successful implementation of the measure. However, participatory methods require considerably more time, additional resources and special training for policy-makers.

A key message is that there is a general shift away from recommending "one-size fits all" solutions. There is a wide and growing base of documented policy experience available in merging efficiency and equity objectives and best-practice examples for a wide variety of institutional and ecological settings. This book discusses a wide range of conceptual and methodological issues, and also uses numerous examples to illustrate how they have been implemented in practice.

PART I

Understanding the Distributional Impacts of Biodiversity Policies

ISBN 978-92-64-03431-0
People and Biodiversity Policies
Impacts, Issues and Strategies for Policy Action

Chapter 1

Introduction

Biodiversity policies are about promoting "the conservation of biological diversity, the sustainable use of its components and the fair and equitable sharing of the benefits arising out of the utilisation of genetic resources" (CBD, 1992). The aim of improving and maintaining biologically diverse habitats and ecosystems is to create net benefits to society by realising all of biodiversity's values: material benefits, as well as those that are less easily quantified.

However, just like any other environmental policy, while biodiversity policies can improve aggregate well-being, they can also create winners and losers (Wells, 1992; Pearce and Moran, 1994; Sunderlin *et al.*, 2005). OECD policy guidelines call explicitly for policy-makers to consider the effects of such policies on the absolute and relative well-being of different groups of people.[1] Thus equity should be as important a policy dimension as environmental effectiveness, administration and compliance costs, and criteria are needed for assessing the distributive effects of environmental policies (Serret and Johnstone, 2006).

Considerable work has recently been done to increase our understanding of the distributive effects of environmental policies (Serret and Johnstone, 2006). The following have been important contributions:

- The development of an overarching framework for assessing the distribution of environmental quality across different groups (Pearce, 2006).
- Determining the financial incidence of environmental policy measures (Kriström, 2006).
- An empirical study of patterns of environmental quality (Brainard *et al.*, 2006).
- Clarifying the link between distributive effects of environmental policies and the thorny issue of environmental justice (Hamilton, 2006).

Many of the insights from this work are relevant in principle to the analysis of the distributive effects of biodiversity policies. For policy-makers charged with developing biodiversity policies, however, this body of research has two drawbacks. The first is that it needs to be made relevant to biodiversity, since little of it refers explicitly to policies aimed at biodiversity. The second is that biodiversity policies differ from other forms of environmental policy in a number of significant ways, most important of which is that much biodiversity policy is inexorably tied to questions of optimal land-use. This leads to distinct policy objectives, a distinct subject matter with distinct policy challenges, a unique set of policy instruments, and an inherently specific geographical scope. The existing literature on distributive impacts of environmental policies is therefore insufficient for informing and providing guidance on biodiversity policies.

The aim of this book is to help policy-makers responsible for designing and implementing biodiversity policies to understand the relevance of distributive issues and how to integrate them into policy formulation. It makes the insights from the theoretical literature on distributive issues in environmental policy accessible and shows how they can inform policy-making in the specific context of biodiversity. Thus, the book provides answers to the following questions:

- Who benefits from and who pays for biodiversity policies?
- How does the choice of policy instrument affect the distribution of benefits and costs?
- How can the distributive effects of these policies be measured, quantified and communicated?
- To what extent should these distributive effects guide the choice between different competing biodiversity policies and what concepts allow a link between policy objectives, distributive outcomes and policy instruments?
- How can policy-makers better integrate distributional concerns into biodiversity policies without compromising conservation goals?

1.1. Study rationale

Specific examples are the best way to illustrate the importance of distributive issues for the design and implementation of biodiversity policies. Three case studies, drawn from different continents and with different policy objectives and instruments, highlight how distributive impacts affect the types of policies that can be pursued, and also the long-term viability of conservation policies.

1.1.1. *Case 1: Lucas vs. South Carolina Coastal Council, USA*

In 1988, the state legislature of South Carolina enacted the Beachfront Management Act following concern about the erosion of unique coastal areas and the economic damage caused by storms. This legislation effectively banned further development in many areas along the state's coastline. The act was to be implemented by the South Carolina Coastal Council (SCCC), and extended an earlier law that had protected coastal areas designated as "critical". The law was also intended to slow the increasing rate of claims for property damage that were being made as a result of storms.

Two years earlier, David H. Lucas had bought two parcels of property on a barrier island to build homes for reselling at a profit. Unfortunately for Lucas, the new act stopped him from completing this project, and because other uses of the land were also not possible, he also lost the nearly USD 1 million investment he had made in buying the parcels. To recover his losses, Lucas filed suit under the Fifth Amendment's "taking" provision of the United States' Constitution.

That amendment had originally been drafted to deal with the confiscation by government of private individual's property for public use. When the role of government was very limited (*e.g.* raising an army, writing laws and enforcing them, protecting commerce) the intrusion of government into economic life was relatively minor and the amendment's scope was limited. The conflict between Lucas and SCCC showed that environmental and biodiversity issues were beginning to have strong redistributive impacts. Lucas claimed that a regulation that left his property untouched but rendered it commercially valueless was akin to physical confiscation. The government argued that the regulation was intended to prevent public harm by avoiding the erosion of the coastline. The court found this argument weak, and instead saw the government's action as conferring a benefit to the general public at the expense of a few individuals. Government has the right to regulate – without compensating any individual – to protect the public from harm. However, when that regulation is predominantly to provide a public benefit, then those individuals who suffer substantial loss should be compensated. The definition of what constitutes a substantial loss is not fixed, but legal tradition puts the reduction in property value at about 65% as a minimum. A few other mitigating circumstances have also been established that can avoid the need for compensation.

Thus the court stated that when there are clear benefits to the public, governments must use their power of taxation to ensure that the costs are shared. The precedent is most relevant to environmental issues such as the protection of biodiversity, since it is difficult to prove that public harm would ensue from a loss of biodiversity. In fact, apart from a few notable exceptions, most environmental amenities are seen by the courts as providing a benefit rather than preventing a harm.

The relevance of this case to distributive issues is that the legal process provides a channel for lessening public reaction to distributive impacts. By requiring government to compensate individuals who are heavily affected by policies that benefit the wider public, it creates a means of venting the strongest negative public responses to biodiversity policy. Thus the kinds of reactions that might derail a policy are deflected. In contrast, as we see in the next case (from Germany), when such channels do not exist or are very difficult to access, there can be a much more widespread backlash against biodiversity policies.

PEOPLE AND BIODIVERSITY POLICIES – ISBN 978-92-64-03431-0 – © OECD 2008

1.1.2. *Case 2: Implementing Natura 2000 in Germany*

In 1992, the European Union (EU) adopted the Directive on the Conservation of Natural Habitats and Wild Fauna and Flora (the "Habitats Directive") to fulfil its obligations under the Rio Summit. This, together with the Directive on the Conservation of Wild Birds ("Birds Directive", 1979), led to the establishment of an ecological network of protected natural and semi-natural habitats called Natura 2000. Each EU member state was obligated to designate areas for inclusion within this network. In a few countries (in particular Germany) the implementation of Natura 2000 caused severe disputes between local residents, especially local landowners, and conservation agencies. In the latter half of the 1990s, citizen protests against the establishment of nature parks in Germany became a frequent feature of the environmental politics scene (see Box 1.1).

Box 1.1. **Opposition to protected areas in Germany**

From the early 1990s, and especially 1995, local, regional and national citizens' groups began to vigorously oppose the establishment of protected areas in Germany, effectively delaying nature conservation efforts, much to the frustration of conservation agencies.

Through various tactics, these groups forced the state governments (Länder) to delay protected area projects. The sometimes hostile nature of local opposition to protected areas manifested itself in public demonstrations and boycotts of public meetings on the protected areas, such as in the Uckermärkische Seen (Uckermark Lakes) Nature Park in Brandenburg. In Brandenburg and other areas, persistent protesters took aggressive measures, such as stealing and destroying nature park signs and deliberately ignoring park regulations. These tactics delayed the designation of the Uckermark Lakes Nature Park by about three years.

In Bavaria, the Prime Minister was met with disapproval when he visited the Bayerischer Wald (Bavarian Forest) National Park in southern Germany. Hundreds of people expressed their displeasure with government plans (both the national park administration and the Bavarian government) to expand the protected area by brandishing placards, heckling and burning ranger equipment. Citizens also opposed the Unteres Odertal (Lower Oder Valley) National Park in eastern Germany near Poland, writing protest letters to ministers of the state government, calling parliamentary parties in the Länder Parliament for their support, and organising public media campaigns. These actions resulted in the head of park management being removed from office.

Source: Stoll-Kleemann, 2001.

These conflicts were prompted by a number of factors (Stoll-Kleemann, 2001):

- Citizens were neither consulted nor involved in decisions to establish nature conservation areas.
- The instruments used to accomplish the conservation outcome were seen as imposing disproportionate and unusual costs on local landowners and residents, both in terms of foregoing productive (agricultural) or consumptive (leisure) uses of the land.
- The designation was commonly perceived as having negative impacts on livelihoods.
- There was only a restrictive compensation policy for those perceived to be negatively affected.[2]
- In countries with a well-developed system of rights and obligations and sophisticated institutions for their monitoring and enforcement, citizens understand that opposition has to come before the projects are implemented. In such systems, failure to oppose a policy is frequently interpreted by courts as tacit consent. Policies that have become a *fait accompli* are therefore much harder to overturn.

The above factors explain why the attempts of German conservation agencies to designate Natura 2000 sites met with local opposition. What were the costs of trying to implement Natura 2000 under these circumstances? They can be summarised in four categories:

1. Cost of delay. The most immediate impact was a delay in protecting the area while other solutions were being investigated or while awaiting the outcome of arbitration and adjudication.
2. Costs of conflict. The most visible costs were destroyed property, the need for additional security forces etc. More significant, however, were the opportunity costs of conflict in terms of time and resources that could more productively have been devoted to other purposes.
3. Costs of policy revision. Many protection policies that have run into opposition have to be extensively revised before undergoing another round of scrutiny.
4. Cost of regulatory risk. Conflict makes it worthwhile for opponents to subject conservation policies to a degree of legal scrutiny that would not otherwise have been required. In the process, higher judicial bodies are frequently called on to define the exact boundaries of policy intervention. As a result, courts can arrive at a much more circumscribed set of policy options than previously thought possible. For example, the Higher Administrative Court in Lüneburg issued a ruling in 1999 questioning whether conservation agencies even had a mandate to declare land as a

PEOPLE AND BIODIVERSITY POLICIES – ISBN 978-92-64-03431-0 – © OECD 2008

part of a protected area if that land had previously been disturbed (Stoll-Kleemann, 2001).

In this book we claim that the distributive issues leading to conflicts surrounding the implementation of Natura 2000 sites in Germany were partially foreseeable. We also argue that these issues must be considered early in the policy process. Finally, we demonstrate that different processes and institutions may achieve conservation objectives that are very similar to those originally envisaged, but at considerably lower aggregate cost to society. It is thus possible to avoid the costs of delay, conflict and policy revision, and the costs of regulatory risk.

1.1.3. *Case 3: Extractive reserves in the Brazilian Amazon*

Since the 1970s, Brazil's national biodiversity strategy has included unique forms of protected areas called "extractive reserves". These are a product of conflict over natural resources and land ownership which pitted indigenous rubber tappers against immigrant farmers (Allegretti, 1990 and 2002). Although the reserves were originally thought of as a proposal for agrarian reform adapted to the needs of populations living from the extraction of forest products, in reality they were conceived as conservation units. The extractive reserves were created to settle this conflict between competing groups of users. These reserves are a highly idiosyncratic mix of property rights assigned to local residents and the government in order to achieve two objectives:

1. Conservation of significant rainforest areas in their original state by banning extractive activities.
2. Economic development of the indigenous population through the reaffirmation and formalisation of extensive usufruct rights (Goeschl and Igliori, 2006).

Table 1.1 lists all the extractive reserves established with these twin objectives in mind.

In many respects, this policy is exemplary: it is broadly benign, with no land being forcibly taken and no resettlements initiated. While initially sluggish, bottom-up involvement of local participants became more effective over time. Also, few direct costs were imposed on the parties. Observers have identified two winners from this conservation policy (Menezes, 1994; Allegretti, 2002): local residents, who gain secure property rights to continue to exploit non-wood forest products; and society at large, which benefits at both the national and global level. This is because the benefits of rainforest conservation in the form of carbon sequestration, watershed protection, maintenance of rare species, etc. are characterised by a significant spatial diffusion, delivering benefits often thousands of miles away. The only

Table 1.1. **Extractive reserves in the Brazilian Amazon**

Name/Federal unit	Area (ha)	Population	Main resources
Alto Jurua – AC	506 186	4 170	Rubber
Chico Mendes – AC	970 570	6 028	Nuts/Copaíba/Rubber
Alto Tarauacá – AC	151 199	–	–
Rio Cajari – AP	481 650	3 283	Nuts/Copaíba oil/Rubber/Açaí fruit
Rio Ouro Preto – RO	204 583	431	Nuts/Copaíba oil/Rubber
Lago do Cunia – RO	52 065	400	Fishery
Extremo Norte do Tocantins – TO	9 280	800	Babaçú fruit/Fishery
Mata Grande – MA	10 450	500	Babaçú fruit/Fishery
Quilombo do Frexal – MA	9 542	900	Babaçú fruit/Fishery
Ciriaco – MA	7 050	1 150	Babaçú fruit
Tapajos Arapiuns – PA	647 610	4 000	Rubber/Fishery/Oil and resin
Medio Jurua – AM	253 226	700	Rubber/Fishery
Total	**3 303 411**	**12 164**	

1. Copaíba is a tree producing oil used for pharmaceutical purposes. Its wood is also used for furniture and construction.
2. Babaçú is a palm. Its nuts are used to produce cooking oil as well as for charcoal and animal feed.
3. Açaí is a palm tree of which both the fruit and the "palm heart" are useful.

Source: Goeschl and Igliori, 2004.

negatively affected groups are those who would have benefited from the conversion of these lands to agricultural use, especially cattle ranching.

At first, the distributional impacts of these extractive reserves resulted in some low-level violence at the interface between expanding agricultural operations and extractive reserves. But these conflicts then escalated, culminating in the murder of the leader of the rubber tapper movement, Chico Mendes. Since then, however, extractive reserves have not suffered from significant open conflict.

Instead, the distributional issues around extractive reserves have moved to much subtler levels. It is now becoming increasingly clear that the development goal inherent in the policy has little realistic possibility of succeeding in generating significant, if any, income gains to the inhabitants of extractive reserves (Southgate; 1998; Goeschl and Igliori, 2004 and 2006). A very partial, but nevertheless telling observation comes from the market for raw natural rubber (Figure 1.1), one of the main output markets for indigenous communities living and operating in these extractive reserves. Due to the restrictions placed on production within the reserves, local producers are unable to match the productivity improvements of their competitors and hence lose out over time to cheaper producers (Homma, 1992).

Implementing biodiversity policies through extractive reserves thus requires local residents to accept that their communities are confined to a low growth path for the foreseeable future by foregoing alternative uses of the

PEOPLE AND BIODIVERSITY POLICIES – ISBN 978-92-64-03431-0 – © OECD 2008

Figure 1.1. **Market share of extractive reserves on raw latex market, Brazil**

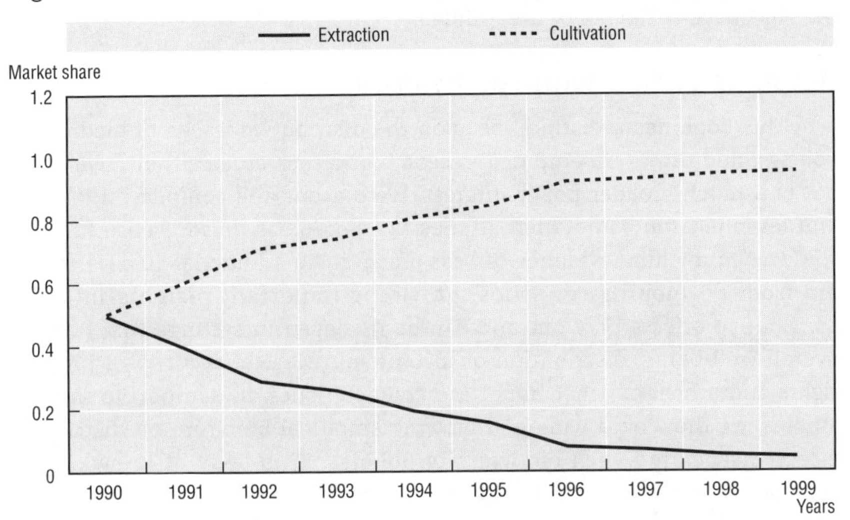

Source: Homma, 1992.

land, while generating considerable benefits to others. This is because the restrictive land use provisions give rise to biodiversity benefits, but have little potential for increasing local well-being (Goeschl and Igliori, 2004). Since the effects of giving up higher income growth become more salient over time, the critical question facing extractive reserves is whether they will remain viable despite the increasing gap between populations inside and outside the reserve. Or will the uneven benefits gradually undermine local people's support for the policy?

These case studies show that distributive issues become salient for biodiversity policy-making in a variety of ways. They demonstrate the importance of the institutional setting for accommodating losers from policies. As shown by the previous two cases (Lucas *vs.* SCCC and Natura 2000 in Germany), where channels for protest are available and can be accessed at reasonable cost, institutions charged with solving distributional conflicts can carry out their work without further delay and without incurring additional cost. Where such channels are not available, distributive issues spill over into other arenas and can be contained only at significant cost in terms of time and resources. As the case of extractive reserves demonstrates, however, even when the short-term distributive issues are broadly resolved, there are important longer-term effects that can threaten or undo the original policy. Policy-makers are therefore increasingly expected to consider the distributive impacts of their policies and to design mechanisms for successfully addressing distributive issues that become policy-relevant while ensuring that the objectives of the policy are fulfilled.

1.2. Objectives and structure of the book

This book has a distinct focus on the distributive issues of biodiversity policies, and draws on empirical research and case studies that have arisen out of a much broader policy interest. We present and compile information and lessons from 44 detailed studies (Annex A). Of these, 34 are in OECD countries, providing evidence of best practice. An additional 10 case studies are from developing countries, providing important insights into how distributive effects play out and can be managed in settings that differ in overall well-being, distribution of income and wealth, security of property rights and a host of other important characteristics. In addition to the case studies, we draw on a wide and growing empirical literature on distributive effects associated with biodiversity policies. This body of knowledge is extensively documented in the list of references at the end of the book.

The book is divided into three parts:

- Part I introduces key concepts in the analysis of distributive impacts of biodiversity policies. It then explains how distributive impacts can be empirically measured, quantified in terms of summary values, and communicated in a policy-making context.
- Part II explores whether distributive issues should be considered within biodiversity-related policies rather than being dealt with separately through the fiscal or other systems of redistribution.
- Part III then describes the main methods for integrating distributive issues into biodiversity policies.

Notes

1. These guidelines originate from the OECD report on *Evaluating Economic Instruments for Environmental Policy* (OECD, 1997).
2. German law distinguishes between two levels of coercive property rights modification when designating protected areas: 1) modification that reduces the value of property deemed to be within what is called the "social obligation of private property"; and 2) modification that reduces the value of property that exceeds that level. Only in the latter case is the modification of property rights deemed to constitute a "taking" (*i.e.* expropriation), which is a prerequisite for owners to be entitled to monetary compensation. An exception can be made for agricultural losses, which are compensable at any level of restriction. This general practice sounds similar to the US, but its implementation has been more difficult, leading to greater resistance to policy.

PEOPLE AND BIODIVERSITY POLICIES – ISBN 978-92-64-03431-0 – © OECD 2008

ISBN 978-92-64-03431-0
People and Biodiversity Policies
Impacts, Issues and Strategies for Policy Action

Chapter 2

Methods for Measuring the Distributive Effects of Biodiversity Policies

For all areas of social policy, the decision to implement a policy should be determined by the balance of benefits and costs. But when we are concerned about well-being, benefits and costs cannot be limited to monetary terms, but must include *any* impact that results from policy implementation. In that sense, there is little disagreement amongst policy analysts: all agree that, broadly, policy must reflect the wishes of the community. Where the disagreement begins, however, is over *how* the benefits and costs will be measured.

In balancing the complete set of impacts, a wide range of issues may need to be considered. This includes direct and indirect impacts from both the concrete and abstract aspects of biodiversity. Those impacts need to be methodically accounted for across many economic, social, spatial and temporal groupings. This chapter discusses how to account for, and measure, those impacts. Given the complexity of measuring impacts, it is essential to have rigorous techniques to guide the analysis. Moreover, it is essential to capture the highest level of detail about the distribution of costs and benefits (of all types).

This is particularly true for distributive issues since it is the details of the distribution of impacts that are the key issues of interest. In fact, in some cases the level of disaggregation directly affects the conclusion of the analysis. More detailed data have sometimes led to conclusions being reversed. This is because impacts that are averaged over a large group can appear to have small consequences for any one individual. But a given impact is much more serious when it affects a small number of individuals very intensely, rather than a larger group more moderately.

Recent years have seen a substantial literature develop to provide empirical estimates of impacts at varying levels of aggregation. The empirical data range from measurement of impacts on individuals within a community to measurement at the global level. The policy impact on individuals or groups can be examined through a variety of measures that attempt to convey its distributive consequences. These techniques range from those that measure changes in income inequality to those that measure changes in employment or child health. In each case there is an underlying valuation being made concerning resulting changes in social well-being. That is, society must have an aversion to increased income inequality in order for it to act to prevent such an outcome from new policies.

 PEOPLE AND BIODIVERSITY POLICIES – ISBN 978-92-64-03431-0 – © OECD 2008

Currently, one of the most important issues concerning biodiversity is the international distribution of costs and benefits. Biodiversity is mostly (though not exclusively) found in developing countries, where income levels are somewhat lower than in most OECD countries. In many cases a significant share of the non-use benefits (see Section 3.2.1) of conserving biodiversity in these countries might accrue to developed countries. This asymmetry is another key dimension of distributive issues at the international level.

Comparing the various techniques used to gauge impacts shows that the applicability of particular methods depends on the policy measure, the geographical scale, and on the data availability. Each of the methods has particular strengths in terms of capturing distributive effects, and weaknesses in terms of either omitting important dimensions or not allowing some type of aggregation to be carried out. This emphasises a key message of the literature on assessing distributive impacts: the method of assessing distributive effects cannot be separated from the policy to be analysed.

2.1. Efficiency, effectiveness and distribution in policy analysis

Biodiversity policies are just one area within the wider ambit of public policy projects. Across all areas of public policy there is the vexing question of what constitutes the "right policy". Over the last 50 years, an important literature on how to evaluate and choose between competing public projects has emerged, under the heading of cost-benefit analysis (CBA). Originally developed for large engineering projects, CBA proved to be a potent tool for policy analysis when combined with welfare economics.[1] In its modern form, CBA provides a rich set of guiding principles, criteria and a methodology to help policy-makers decide between policies. Although alternative policy-decision approaches exist, none is as widely used as CBA.

It is not possible here to give a full account of the relationship between welfare economics and CBA (for excellent treatments see Pearce, 1983; Hanley and Spash, 1993; Just *et al.*, 2004). What is important is that one of CBA's key contributions is to allow the policy analyst to use two useful concepts from welfare economics in making policy choices: *efficiency*, and the closely related concept of *(cost) effectiveness*. In its most demanding form (Pareto efficiency – see Box 5.1 in Chapter 5 for more), efficiency refers to a situation in which it is not possible for a policy to improve the well-being of at least one member of society without reducing the well-being of some other. Cost effectiveness is a less demanding criterion, requiring a policy to accomplish a certain objective with a minimum sacrifice by society.

Through the pervasive use of CBA in applied policy analysis, efficiency has taken on a somewhat different meaning in policy-making. For policy choice to be *efficient* requires that *the policy chosen should maximise the difference between the benefits delivered by the policy and the cost of implementing the policy*. A policy is efficient if no other policy can improve the social surplus (*i.e.* the net improvements in welfare aggregated across all individuals of relevance to the policy-maker). Both notions build directly on the idea of the *social surplus* and its benefits (seen as gains in utility) and costs (seen as loss in utility). We use this notion of efficiency most often in this book. Since the policy-maker would, in theory, have to be informed about all relevant costs and benefits of each project and would have to be aware of all potential projects in order to call a policy efficient, (cost) effectiveness and efficiency are in practice often closely linked.

The abstraction of the efficiency criterion in policy choice from the distributive consequences of these policies is the starting point for this book. While both the fundamental intellectual appeal of this view of efficiency and its usefulness for informing policy choice are not in dispute, there is a concern based on empirical observation and a shift in policy foci about whether the distributional consequences of policies should be accorded greater weight and thus give rise to "better" policies, in a sense that will become clear later. While this concern for distributional consequences is not specific to biodiversity policies (see *e.g.* Serret and Johnstone, 2006), what is specific is how biodiversity policies give rise to these distributive effects.

2.1.1. Biodiversity, cost benefit analysis and efficiency

Recent years have seen tremendous progress in translating the complex resource of biodiversity into the framework of CBA (Pearce and Turner, 1990; Pearce and Moran, 1994, Perrings *et al.*, 1995, Swanson, 1995; Costanza *et al.*, 1997; Dasgupta, 2000). "Making biodiversity count" in assessing the benefits and costs of policies has enabled policy-makers to advocate for resources to be spent on the maintenance and improvement of biologically-diverse habitats and ecosystems. The OECD in particular has developed a range of documents to help policy-makers understand how biodiversity conservation is an activity whose benefits can be measured, incorporated and communicated using CBA (OECD, 1996; OECD, 2002; OECD, 2003). Given the competition between different policies for resources and funds, this progress has had two important results. The first is a general awareness among environmental policy-makers that many of the externalities associated with biodiversity are worth addressing through policies since they can be demonstrated within a CBA framework to further society's welfare. The second result is to demonstrate that biodiversity policies can pass the efficiency test, giving the same robust foundation to arguments for maintaining and enhancing habitats and ecosystems as those for addressing other policy issues.

PEOPLE AND BIODIVERSITY POLICIES – ISBN 978-92-64-03431-0 – © OECD 2008

The analysis that follows builds on this important work and uses it to understand the various patterns of winners and losers that the pursuit of biodiversity policies generates across society at different spatial and intertemporal scales. The analysis also demonstrates the continued overwhelming relevance of efficiency considerations for biodiversity policy-making, but highlights that there are circumstances in which distributive effects should play a role. The most important of these are because addressing distributive issues helps policies to reflect a more profound notion of efficiency than has been present in welfare economics to date, but that can be overlooked in a mechanical application of CBA for policy decisions.

An analysis of the distributive effects of biodiversity policies is also necessary now that policy-makers are under increased pressure to demonstrate that their policies are informed by and comply with criteria emanating from global policy discourses such as the Millennium Development Goals. These criteria frequently contain explicit distributive objectives with little or no guidance on how to fit biodiversity policies around them. A key message of this book is that considerations such as poverty eradication or benefit sharing will pose important challenges to policy design and instrument choice if a focus on efficiency is to be retained.

2.2. Empirical measures of distributive effects

The first step in the assessment and possible mitigation of distributive impacts of biodiversity policies is identification and measurement. A "baseline" needs to be established against which action/inaction will be measured. Any foreseen impacts of the policy can be compared to the baseline, to determine their extent on distributive factors. Subsequent choices concern the level of measurement and the specific representation of distributional effects in the form of some measure. Modern techniques offer a wider range of possible measures for capturing distributive effects.

2.2.1. Levels of measurement

Distributive effects can be measured at very different levels of aggregation. Commonly used base units for distributive analysis are individuals, households, families, communities, groups, industries, regions or countries.

The appropriate level of aggregation depends on the specific policy context. Two objectives need to be balanced: *i)* the need for an exhaustive survey of all distributive effects of a policy measure (meeting this objective requires analysis at the individual level); *versus ii)* the need to succinctly summarise, understand and communicate distributive effects for policy purposes (this usually calls for higher levels of aggregation).

In the practice of policy analysis, therefore, policy-makers are often, and for good reason, presented with distributive impacts measured at highly aggregated levels. Distributional data in this book will also be frequently given at the household, group or industry level. However, the reader of such summary information should remember that a high degree of heterogeneity may be hidden at such a level of analysis. From an economic welfare perspective, it is expected that a policy will have non-proportional impacts when it applies to a small group instead of a large group.[2] Similarly, reporting aggregated numbers may mask the impact of a policy on a small group. Thus the level of aggregation of a policy, and its reporting, can have substantial impact on the results. The wider literature on intra-household distribution of income and wealth is instructive in this regard.

2.2.2. *Measuring aggregate gains: distributive effects of biodiversity policies*

Since biodiversity policies must be considered against the backdrop of other issues that governments address – and trade-offs are often needed across agendas – governments need metrics that can compare magnitudes. CBA is the most common method for measuring the trade-offs between policy issues (Pearce *et al.*, 2006). What are CBA's characteristics when looking at the distributive impacts of a particular policy design?

- It measures costs and benefits in monetary terms, and thus captures some impacts on stakeholders. Nevertheless some important effects cannot be measured in monetary terms, so other complementary methods are required.
- It may have difficulty dealing with the distribution of costs and benefits at different geographical scales (local, regional, global) and among stakeholder groups.
- It focuses on economic aspects, and thus may require the addition of other factors (*e.g.* social criteria) into the decision-making process.

In some settings, CBA can make distributive effects explicit without considerable adjustments to methodology or data requirements. In the next chapter, a simple extension of CBA is applied to the analysis of distributive impacts.

Biodiversity policies, like other interventions, affect different groups differently. What tools can capture this distribution in a meaningful way?

- Methods based on income-equivalent measures.
  - ❖ **Summary measures of (in)equality:** *e.g.* measures of income or wealth inequality, such as the Lorenz curve and the Gini coefficient.

- **Extended versions of CBA:** *e.g.* the calculation of costs and benefits by different stakeholder groups at different geographical scales – CBA with distributional matrices.
- **Social accounting matrix (SAM):** offers an enhanced representation of economic and social effects.
- **Distributive weights:** take into account that a person's welfare gain derived from receiving an additional unit of income typically decreases as income goes up (diminishing marginal utility of income).
- **Atkinson inequality index:** integrates normative judgements about welfare into the distributive analysis.

- Alternative measures:
  - **Employment-based analysis:** distributive outcomes are measured by reference to the ability of affected groups to generate income out of employment.
  - **Child health-based analysis:** uses the measures of the changes in nutritional status of children as a summary indicator of welfare impacts of policies.
- Multidimensional measures:
  - **Stochastic dominance analysis:** a multidimensional approach to welfare that is able to rank policy outcomes simultaneously across several criteria such as income, inequality and poverty using statistical techniques.
  - **Multi-criteria analysis (MCA):** seeks to integrate social and cultural aspects into the analysis.
  - **Social impact assessment** with a stakeholder analysis: records the interests and attitudes of the stakeholder groups alongside conventional assessments of positive and negative effects of the proposed policy on the groups.

The first class of methods (income-equivalent measures) is common in the economic analysis of distributive impacts. These emphasise quantifiable effects and prefer to condense information. The remaining methods are currently popular in the social policy evaluation literature. They combine quantitative and qualitative data to capture some of the complexity of distributive impacts beyond their economic dimension. It is important to understand that the different methods have different data requirements. Therefore, while it may be desirable in an exhaustive analysis of distributive effects to use several measures, extending the number of measures and dimensions assessed requires additional time and resources.

Below we describe these measures of distributive effects in more detail, followed by examples of their application in an actual policy assessment context. We then compare and contrast the different methods (see Table 2.15).

2.3. Methods based on income-equivalent measures

2.3.1. *Lorenz curves, Gini coefficients and other economic measures of inequality*

A classic measure of inequality in economics is the concentration curve or Lorenz curve (Figure 2.1), named after Max Lorenz, its proponent. This curve is a versatile tool for measuring the distribution of a single-dimensional quantitative characteristic, *e.g.* income and wealth, in a population. Its most common use is in economic studies of income distribution.

Figure 2.1. **Lorenz curve**

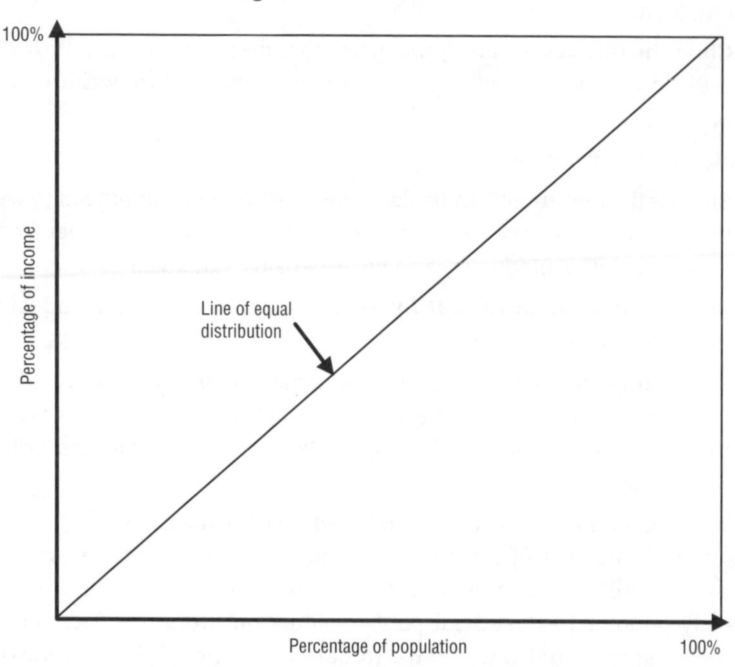

Graphically, the Lorenz curve is displayed as a curve in a two-dimensional diagram (see Figure 2.1). In this example, a point on the Lorenz curve indicates what share of society's total available volume of income the poorest *x*% of households or individuals receives. Taking income as an example, the horizontal axis measures the percentage of the population, starting with the households or individuals at the bottom of the income distribution, and the

 PEOPLE AND BIODIVERSITY POLICIES – ISBN 978-92-64-03431-0 – © OECD 2008

vertical axis measures the percentage of the total income, again starting with the households or individuals at the bottom of the income distribution. If the Lorenz curve of the distribution of the asset is a 45° line, then income is perfectly equally distributed. The greater the deviation of the curve from the 45° line, the more uneven the distribution. A simple measure of this deviation is the Gini coefficient. A Gini coefficient of zero indicates a perfectly even distribution; a Gini coefficient of one indicates a perfectly uneven distribution (implying that one individual or household receives the entire asset). The Gini coefficient has a long history of use in studies of income inequality. In the environmental context, it has been used to analyse the distribution of pollution in the US (Millimet and Slottje, 2000).

The Lorenz curve has low data requirements and offers a succinct and graphically appealing representation. It is closely related to other economic measures for analysing the distribution of income and wealth.

Example: Measures of equality of income distribution of privatisation of mangroves in Vietnam (Adger et al., *1997)*

This examples explores how the conversion of common property mangroves for private agriculture and aquaculture in the Quang Ninh Province of Vietnam affected the equality of income distribution of two villages: Le Loi and Thong Nhat Communes. Using the Gini coefficient and breaking down total income into its constituent sources by origin (farming, fishing, commerce, and outside income), the authors identified which activities give rise to greater unevenness in income distribution. Table 2.1 shows how different income sources contributed to inequality in surveyed households in Le Loi and Thong Nhat.

Table 2.1. **Contribution of income sources to inequality**

Income sources	Gini/pseudo Gini	Share of income %	Contribution to inequality	Inequalising effect
Overall income	0.436			
Farming income	0.351	66.9	54.6	–
Fishing income	0.334	4.7	3.6	–
Commercial activities	0.692	3.8	6.0	+
Wages, transfers and remittances	0.624	24.6	35.7	+

Source: Adger *et al.*, 1997.

In this example, the share of income is the percentage of total income derived from each source. The contribution to inequality of each income source is measured as the percentage of the Gini coefficient attributable to this source relative to its share in income. This allowed the authors to identify that commercial activities and outside income (such as wages and transfers) were causing inequality since they contribute more to inequality than their share of income. The authors concluded that in the mangrove areas, the shift from common resource management to private agriculture and aquaculture increases inequality. In this case, therefore, common property regimes appear to support sustainable natural resource management and to maintain an even income distribution (Adger *et al.*, 1997).

2.3.2. Extended CBA

Classical CBA calculates costs and benefits for a certain time horizon, subtracts costs from benefits and discounts the net amount (*i.e.*, calculates the net present value, NPV). In an extended CBA the stakeholders are also taken into account by making explicit which group has generated the costs and benefits. The calculation of extended CBA thus shows which groups are gaining and which are losing under different scenarios (Table 2.2).

Table 2.2. **Extended CBA by stakeholder group**

	Scenario 1	Scenario 2	Scenario 3
Costs			
Stakeholder 1			
Stakeholder 2			
Stakeholder 3			
Benefits			
Stakeholder 1			
Stakeholder 2			
Stakeholder 3			
NPV			

When the extent of a proposed policy is limited (*i.e.* targeted at a well-defined group), and the secondary consequences are small, then monetising the main costs on the affected group can provide a good approximation of the welfare effects. Examples include restrictions on farming activities or reintroduction of predator species in areas where they may cause damage to livestock. In these cases opportunity costs and/or damage costs will be important to calculate for anticipating the response to the policy and the level of loss that may have to be compensated.

Table 2.3 (used in extended CBAs by the UK Treasury) illustrates the basic core information needed for determining how the costs and benefits of an option are spread across different income groups. When the intention is to minimise the impact on lower income groups, various measures can be undertaken to fill in the table and this information can then be used to rate

PEOPLE AND BIODIVERSITY POLICIES – ISBN 978-92-64-03431-0 – © OECD 2008

Table 2.3. **Income ranges by quintile of equalised net income**

Amount per week	Single with no children	Couple with no children	Single with child	Couple with child	Single with two children	Couple with two children	Single pensioner	Pensioner couple
Quintile of equalised net income								
1								
2								
3								
4								
5								

Source: HM Treasury, 2003.

more favourably those measures that provide greater benefits to lower income quintiles.[3]

Example: Benefits, costs and their distribution among stakeholders in Ream National Park, Cambodia (de Lopez, 2003)

This example describes an economic analysis of Ream National Park, Cambodia. The aim was to assess the benefits and costs of three potential park scenarios and their distribution among stakeholders. The three scenarios were: 1) Experimental park: the current level of forest protection would be maintained, but there would be no protection of fisheries, leading to their eventual collapse; 2) Ghost park: no protection of forests and fisheries such that all timber and fish would be harvested, destroying the area; 3) Dream park: full protection of resources, with only subsistence activities, recreation, education and research permitted. A household survey of local communities provided social, economic and ecological data. The following benefits were assessed: *a)* monetised benefits: non-timber forest products, marine and fresh water products, timber from evergreen forests and mangroves, recreation and tourism, protection from storm and erosion; *b)* non-monetised benefits: marine ecosystems, medical resources, carbon storage, protection from saline water, education and research, culture, option value and existence value.

Table 2.4 shows the net present values of the three management scenarios. Protection scenarios (options 1 and 3) allocate the bulk of the park's benefits to local communities. The dream park scenario confers three times more benefits to the villagers than the ghost park. In the latter scenario, however, local communities whose traditional livelihoods depend on the sustainable use of the park would lose the most, while commercial loggers and fishing fleets, as well as the armed forces, would gain most from the exploitation of timber and marine resources.

Table 2.4. **Net present values of different management scenarios**

Benefits and costs	Scenario 1. Experimental park (PV USD)	Scenario 2. Ghost Park (PV USD)	Scenario 3. Dream park (PV USD)
	Benefits		
Wood from mangroves		572 716	
Wood from non-mangroves		5 842 761	
Firewood	853 688		853 688
Fencing	180 232		180 232
Food	134 804		134 804
Roofing	102 061		102 061
Medicine	82 181		82 181
Fisheries	5 207 267	3 576 067	7 867 328
Recreation	21 390		699 636
Protection from storm and erosion: houses	2 605 037		2 605 037
Protection from storm and erosion: crops	539 069		539 069
Protection from storm and erosion: animals	299 376		299 376
	Costs		
Park management costs	255 407		851 356
Capital investment			379 079
Total NPV at 10% discount rate	**9 765 845**	**9 991 544**	**11 896 705**

Source: de Lopez, 2003.

This case shows that an extended CBA can assess the distributive issues well when the main stakeholder groups are identified, and can also identify the benefits that accrue to each group.

2.3.3. Social accounting matrix

The social accounting matrix (SAM) is an analytical framework in which social and economic data are integrated and harmonised, as in an input-output matrix, but expanded to include owners and factors of production and their expenditures. It is constructed as a square matrix (see Table 2.5 for a simplified example), which brings together data on production and income generation by different institutional groups and classes, and data about expenditure of these incomes (OECD, 2003).

It was initially developed at a national level, but is now also used to analyse regional and local economies. It is also a very valuable tool to fully understand the direct and indirect effects of particular interventions in a village or small community (Taylor and Adelman, 1996) by tracking the flow of income received by a sector from other sectors (reading along the rows) and

Table 2.5. **Example of a social accounting matrix**

	Suppliers	Households	Government	Rest of the world
Suppliers				
Households				
Government				
Rest of the world				

the flow of expenditures of a sector to other sectors (reading down the columns).

SAM models have been successfully used to model the economy-wide impacts of sectoral changes; for example, the distributive effects of changes in forest resource policies (Alavalapati *et al.*, 1999).

Example: Distributional impacts of alternative forest management of the Upper Great Lakes region, USA (Marcouiller and Stier, 1996)

A social accounting matrix (SAM) was developed to investigate the distributional impacts and different outputs of alternative forest management practices in the USA's Upper Great Lakes forested region. Different forms of forest management would emphasise different outputs from the forest, ranging from timber to recreational uses such as tourism. These would have different outcomes in both income distribution and output trade-offs. A SAM that includes non-market assets (or public goods) created by the existence of the forest itself offers a method for incorporating the role of these assets in determining income and production. A proportion of the SAM is given below as an illustration (Table 2.6), with the public goods aspect of forest management included as a fourth production factor and contributing to the retail/service sector (sector 6) in the form of recreational services.

The calculation showed that at the time of assessment (1993), growing trees (timber production) and wood processing represented roughly USD 264 million and USD 9.38 billion respectively in the Upper Great Lake States. Total household incomes were roughly USD 47 billion. Calculated using the opportunity cost approach, public goods from forest lands totalled close to USD 80 million.

Numerous strategies are available to increase the supply of public goods. The approach that is modelled in the example is a shift to uneven-aged selective silvicultural techniques. Using the SAM, the increases in income generated by such an increase can then be traced back to recipients. This is done using information on ownership of production sectors.

Table 2.7 illustrates the effects on regional households of a USD 100 million increase in public good output. This shows that such an

Table 2.6. **Part of the environmental SAM for 101 counties within the forested portion of the Lake States**

USD millions

	Production sectors							
	1	2	3	4	5	6	7	8
Production sectors								
1. Agricultural production	223.3	9.5	58.7	1 133	21.5	26.3	13.3	2.4
2. Timber production and services	57.1	20.8	0.7	0.3	171.4	0	0	0
3. Manufacturing	195	19.4	4 804.1	224.5	1 093.3	1 369.3	842.3	417.7
4. Food/fibre processing	13.3	5.9	11.7	975.8	2.1	261.3	0	9.1
5. Wood processing	11.1	0.2	551.4	91.9	1 087.2	30.2	1.1	4.6
6. Retail/services	96.1	5.2	2 779.9	233.3	673.2	1 183.2	409.2	148.7
7. FIRE	90.1	12.9	770	26.4	110.1	609.4	934.2	102.5
8. Government	2.9	0.5	125.1	7.9	42.8	104.3	113.4	35.2
Factor accounts								
1. Labour	441.7	42.2	8 091.7	579.8	2 207.2	10 717.7	1 343.3	7 084.6
2. Capital	344.3	31.4	5 339.8	521.4	1 342.4	4 683.6	3 662.4	62.4
3. Land	585.8	35.5	0	0	0	0	0	0
4. Non-market assets	0	0	0	0		77.6	0	0

Source: Marcouiller and Stier, 1996.

Table 2.7. **Impacts on regional households**

Account	Initial household income distribution		Fixed price impact	
	Million USD	%	Million USD	%
Household				
Low income	8 847	18.9	7.84	8.9
Medium income	21 415	45.6	49.44	55.9
High income	16 655	35.5	31.14	35.2
Total	46 917	100.0	88.42	100.0

Source: Marcouiller and Stier, 1996.

increase would favour medium and high-income households disproportionately over low-income households.

2.3.4. Distributional weights

The extended CBA and the social accounting matrix make the distributional impact of policies transparent to the decision-maker. However, they do not provide any guidance on how distributional issues should affect

 PEOPLE AND BIODIVERSITY POLICIES – ISBN 978-92-64-03431-0 –

policy decisions. Assigning distributional weights (Box 2.1), if coupled with one of the former methods, can provide such guidance (Pearce, 1998).

Box 2.1. **The economic theory behind distributional weights**

Empirical evidence shows that some rough rules-of-thumb provide plausible grounds for assigning weights to different income groups (*e.g.* quintiles). In these calculations a benefit, or cost, accruing to a relatively low-income family would be weighted more than one accruing to a high-income family (HM Treasury, 2003). Economics thus offers a straightforward way of correcting for an implicit bias towards the rich. Since the functional relation between income and the marginal value of consumption can be reasonably well estimated, one can supplement plain CBA with "distributional weights" in order to correct for the wealth bias (Drèze and Stern, 1987; Drèze, 1998; Johansson-Stenman, 2005). The augmented linear policy model (Figure 5.2 in Chapter 5) shows these weights as modifying the measurement of welfare impacts.

But what exactly are these weights, how are they calculated and applied and how do they affect the choice of the optimal biodiversity policy? What is needed to calculate the weights is a functional relationship between consumption, c, and utility, u, *i.e.* the individual's utility function $u(c)$, or at least its first derivative, $u'(c)$. The purpose of distributional weights is then to make different levels of consumption c comparable. A reasonably simplified application based on empirical estimates can be found in the *UK Green Book* (HM Treasury, 2003). It assumes individuals have utility that can be represented by:

$$u(c) = In\ c$$

which corresponds to a marginal utility of consumption of 1/c. Hence, if one knows an individual's income, m_i, which is a reasonably good approximation of his or her consumption, one can calculate his/her marginal utility. The marginal utility of consumption is strictly decreasing in income. To get the distributional weight one has to express an individual's marginal utility as a percentage of average marginal utility. The *UK Green Book* uses median income $\overline{m}$ to calculate the latter. The distributional weight w_i of individual i is therefore:

$$\alpha_i = \frac{\overline{m}}{m_i}$$

Distributional weights are clearly decreasing in income. Hence, by multiplying an individual's mean willingness to pay (MWTP) with the corresponding distributional weight before aggregation, one counteracts the wealth bias. The evaluation of public policies is no longer systematically regressive.

Consider two options for implementing a specific biodiversity policy. The policy will have a negative impact on one of two stakeholder groups. Table 2.8 displays the results of a study to determine these groups' willingness to pay (MWTP) to avoid these impacts. The assumed median income is USD 40 000.

Table 2.8. **Two options for implementing a specific biodiversity policy**

	Group	Aggregated MWTP to avoid impacts	Size of group	MWTP of individual	Annual individual income	Distributional weight	Distribution-adjusted cost of policy
Policy 1	Poor	USD 10 000	1 000	USD 10	USD 20 000	2	USD 20 000
Policy 2	Rich	USD 50 000	1 000	USD 50	USD 200 000	0.2	USD 10 000

This stylised example illustrates how important distributional concerns can be in policy evaluation. A plain CBA would compare the two groups for their aggregated MWTP to avoid the negative effects. For the "poor" option, the MWTP to avoid impacts is USD 10 000, as opposed to USD 50 000 for the rich option (column 3). Dollar-for-dollar, therefore, the social cost of imposing the policy on the "poor" is significantly less than that of imposing it on the "rich". Based on the plain CBA, the first option would have to be recommended, *i.e.* the one that places the burden of the policy on the poor. Augmenting the CBA with distributive weights (column 7), however, leads to a different conclusion. This is because the individual MWTPs are adjusted by appropriate distributional weights that take into account that the sacrifice of USD 10 at an income of USD 20 000 should weigh heavier than the sacrifice of USD 50 at an income ten times higher. Applying this logic to the aggregate "MWTP to avoid" (along the lines of the *UK Green Book*, HM Treasury, 2003) leads to the conclusion that the social cost of the first option should be calculated as twice that of the second. The augmented CBA would therefore recommend imposing the cost on the rich in order to minimise the social costs of achieving the biodiversity objective.

2.3.5. *Atkinson inequality index*

An advanced economic measure of inequality is the Atkinson index. Unlike other measures of inequality, such as distributional weights, it explicitly embodies normative judgements about social welfare (Atkinson, 1970). It is the most prominent example of "equivalence scales" in the analysis of income and wealth (Atkinson and Bourguignon, 1982). The index is derived from the equity-sensitive average income (y_e). This is defined as the level of per capita income which – if enjoyed by everybody – would make total welfare exactly equal to the total welfare generated by the actual income

 PEOPLE AND BIODIVERSITY POLICIES – ISBN 978-92-64-03431-0 – © OECD 2008

distribution (see Box 2.2). The y_e is first calculated to give a redistribution of income that is consistent with a measured value of people's preferences for equality without any loss of total welfare. That measure is then compared to actual average income to give an index of inequality that is counter to people's preferences. In other words, people's preference for equality is measured and then used to determine how far actual income distribution is from the preferred distribution.

Box 2.2. **The equity-sensitive average income**

Formally, the equity-sensitive average income is given by:

$$y_e = \left(\sum_{i=1}^{n} f(y_i) y_i^{1-e} \right)^{\frac{1}{1-e}}$$

Where y_i is the proportion of total income earned by the ith group, and *e* is the so-called inequality aversion parameter. This parameter summarises society's preferences for equality, and can take values ranging from zero to infinity. If $e > 0$, there is a social preference for equality (or an aversion to inequality). An increase in *e* is associated with a stronger social preference for income transfers at the lower end of the distribution and lower preference to transfers at the upper end. Typically used values of *e* include 0.5 and 2, but empirically calibrated measures are available based on experimental evidence (Amiel *et al.*, 1999).

The Atkinson index (I) is then given by:

$$I = 1 - \frac{y_e}{\mu}$$

where μ is the actual mean income. The more equal the income distribution, the closer y_e will be to μ, and the lower the value of the Atkinson index. For any income distribution, the value of I lies between 0 and 1. Like other indices, the Atkinson index is sensitive to the presence of a greater share of the population at the lower end of the income distribution.

The Atkinson index has not received the attention of economists outside the narrow field of income studies. Its applicability to the environmental or biodiversity contexts is therefore an open question.

2.4. Alternative one-dimensional measures

The methods described above use income-equivalent measures to evaluate a policy. However, in some situations it might not be possible or desirable to monetise the relevant policy impacts. In such cases alternative measures can be used as proxies for the welfare effect of a reform.

2.4.1. *Employment-based analysis*

Employment-based analysis (EBA) measures the number of jobs provided or the number of families supported by alternative economic activities using a given resource (typically land) over a given period of time. It can be used to measure the distributional effects of a certain policy or different projects, especially in less developed regions (Taylor, 2001). The rationale underlying the use of EBA is that employment is a constituent component of welfare and determines wider measures of opportunity (Sen, 1997).

EBA is very relevant to biodiversity policies. In previous reports, the OECD has highlighted the potential trade-offs involved between environmental and employment policies in developed economies (OECD, 2003). Also, in the context of species protection programmes, conflicts frequently arise out of a perceived trade-off that pits jobs against conservation (Freudenburg *et al.*, 1998; Meyer, 2001).

Example: The comparison of employment-based and cost-benefit analyses in Yucatán, Mexico (Taylor, 2001)

This analysis compared alternative agricultural activities (traditional farming, improved farming and cattle ranching) using employment-based analysis (EBA) and cost-benefit analysis (CBA). The results indicate that while cattle ranching has the highest benefit-cost ratio (far right column of Table 2.9), both traditional and improved farming provide the greatest employment-based benefits (centre column).

Table 2.9. **Employment-based analysis**

	EBA, people-years for 100 ha, 20 years	CBA, benefit-cost ratio
Traditional farming (milpa[1])	641	–
Improved farming (milpa)	1 495.6	1.54
Cattle ranching	109.6	2.12

1. Milpa refers to a traditional Mesoamerican growing system.
Source: Taylor, 2001.

The example shows that EBA can add value to CBA because it captures highly relevant social effects by calculating the impact on employment of different scenarios. This can also be an important tool for predicting conflict, in particular when combined with empirically validated data from comparable policy settings. Care has to be taken, however, that counting jobs may hide

 PEOPLE AND BIODIVERSITY POLICIES – ISBN 978-92-64-03431-0 – © OECD 2008

important dimensions of job status, qualification match and labour market frictions (Freudenburg *et al.*, 1999).

2.4.2. Child health-based analysis

The literature on multidimensional analysis (see below) emphasises the importance of unconventional measures of welfare, especially if more traditional measures like monetary income are hard to measure or play only a minor role in overall household income. One example of a non-traditional measure of human welfare in the context of biodiversity policies is the nutritional status of children (Gjertsen, 2005). The nutritional status of children in developing countries depends on a number of factors, most critically parental financial income, family harvesting activities from natural assets and government expenditures on child health. It also incorporates a strong intertemporal dimension given the correlation between child nutritional status and lifetime well-being.

Example: Marine protected areas and nutritional status of infants in the Philippines (Gjertsen, 2005)

Gjertsen (2005) used village-level changes in the proportion of underweight children as an indirect measure of welfare impacts of biodiversity policies. Since malnutrition is common in the study area, and both food and monetary income mainly stem from fish provided by degraded coral reefs, there is a direct link between biodiversity conservation and children's nutritional status. A quarterly weighing programme of preschool children provides a much more detailed database than for monetary household income.

A conservation policy to protect degraded coral reefs (Marine Protected Environments, MPA) restricts fishing in certain areas. A win-win situation could arise if fish stocks inside the MPAs increase sufficiently to allow catches in neighbouring areas to outweigh the losses incurred by restrictions imposed by the conservation policy. On the other hand, lose-lose situations could arise if there was no improvement, or even a decline, in fish stocks as a result of the MPA at the same time as fishing opportunities were being foregone by the local population. The empirical analysis failed to find any design variables for the MPA that generated short to medium-term win-win outcomes between protection and child health. However, it did identify design variables that enhanced protection without compromising children's nutritional status, opening up the possibility of long-term win-win outcomes.

While suffering from various statistical problems (such as endogeneity, statistical insignificance of coefficients, and omitted factors), the analysis nonetheless demonstrates the potential value of unconventional measures for analysing distributive issues.

2.5. Multidimensional measures

The previous measures implicitly assume that all relevant impacts can be aggregated into a one-dimensional scale. If, however, some aspects are, at least from the perspective of a central decision-maker, incommensurable, then a multidimensional approach should be adopted.

2.5.1. Stochastic dominance analysis

Pioneered by Kolm (1977), multidimensional analysis is a recent development in the economic analysis of inequality that addresses two criticisms of the conventional economic approaches of the Lorenz curve and the Gini coefficient. One is that welfare of individuals and households will in many circumstances be determined by non-monetised assets such as health, education and crime, and that these are not adequately reflected (Sen, 1997). The other is that the distribution of income and wealth is not determined by the same factors as those determining the distribution of other dimensions of welfare. Income and wealth are therefore poor proxies for other dimensions of human welfare (Justino *et al.*, 2004). Many of these concerns about the multidimensionality of welfare, and by implication of inequality, are widely shared in economics.

The criticism of conventional single-dimensional measures has given rise to attempts to construct measures of multidimensional inequality that can compare distributional outcomes across several aspects of human welfare (*e.g.* income and education). One of these approaches is used in the groundbreaking 1990 United Nations *Human Development Report*, which broadened its determination of welfare to include measures of child mortality, health status, education, etc. (UNDP, 1990).

Comparing and ranking outcomes rather than just putting them side-by-side, however, requires more sophisticated methods. The most established of these uses stochastic dominance criteria to derive statements about multidimensional inequality (Atkinson and Bourguignon, 1982; Maasoumi, 1986). In order to state that one multidimensional distributive outcome is more or less inequitable than another, strong assumptions are necessary about how these outcomes relate to welfare at an individual level (Atkinson and Bourguignon, 1982). To take this analysis to the level of direct policy recommendations requires explicit specification of social weights attached to the different constituent dimensions of welfare (Trannoy, 2003).

PEOPLE AND BIODIVERSITY POLICIES – ISBN 978-92-64-03431-0 – © OECD 2008

Farming versus *fishing in Bangladesh (Islam and Braden, 2006)*

Floodplain management is a key issue in Bangladesh, where both fishing and agriculture are important sources of income. Additionally, flooding is a random event, bringing issues of risk to the fore in decision-making. Agriculture benefits from active floodplain management because managed irrigation and nutrient flows increase agricultural production. However, some floodplain management measures have a negative effect on the catch of fish. This is important since fish are a source of subsistence to the local poor. This means that there are complex trade-offs between farming and fishing that need to be taken into account in floodplain management decisions. Islam and Braden (2006) used stochastic dominance analysis to rank different floodplain management policies, such as varying the height of the floodplain embankment. Higher embankments increase the agricultural benefits, but decrease fish catch (Table 2.10). First-degree stochastic dominance (FSD) identifies the no-embankment scenario as dominating all three embankment scenarios. Thus, the optimal floodplain management strategy might actually consist of no embankment at all. With more information on utility functions of the affected population, second-degree stochastic dominance could then be used to rank the embankment scenarios, which cannot be done completely on the basis of first-degree stochastic dominance.

Table 2.10. **Mean returns of alternative management scenarios and stochastic dominance**

Management scenario	Net returns (million BDT)		Degree of stochastic dominance
	Mean	Std. dev	
Base – no embankment	5 810.64	260.62	FSD over all other scenarios
Low embankment	5 007.63	255.87	FSD over medium embankment FSD by base model
Medium embankment	4 968.94	265.84	FSD by base model
High embankment	5 195.62	297.28	FSD by base model

Source: Islam and Braden, 2006. Returns are measured in 2000 BDT, Bangladeshi taka.

2.5.2. *Multi-criteria analysis*

Multi-criteria analysis (MCA) can be used to compare alternative policy or project scenarios along a set of criteria. The criteria can be measured in monetary or non-monetary units, and even in qualitative terms. Decision-makers (sometimes with the involvement of stakeholders) can assign weights to the different criteria and calculate the best scenario (Nijkamp *et al.*, 1990). Assigning weights can be a lengthy and controversial process, but when it is successfully completed it can make the use of MCA much more effective, since it allows the issue to collapse into a single dimension. A fundamental difference between this approach and other multidimensional measures is that the weights can be assigned by the stakeholders directly.

Distributive effects can be included in the MCA as one or more criteria (Table 2.11), for example, increase in employment, access of certain groups to natural resources, change in the income level of certain stakeholder groups or regions, or global and local benefits. The weights assigned to the social criteria will determine how strongly social and distributive aspects will be taken into account in the decision-making process.

Table 2.11. **Multi-criteria impact matrix**

Criteria	Scenario		
	A	B	C
Economic 1			
Economic 2			
Social 1			
Social 2			
Ecological 1			
Ecological 2			

Example 1: Multi-criteria analysis in Australian forestry policy in New South Wales (Proctor, 2000)

Of the 157 million hectares of forest in Australia, around 20% are comprised of rainforest and open eucalypt forests, both of which are major sources of timber production in the country. Australian governments have carried out comprehensive regional assessments to address the wide variety of different forest values, including biological diversity, areas of wilderness and old growth, cultural, indigenous and heritage values and social and economic implications. They are designed to determine the major part of the Australian forest policy for the next 20 years. These assessments have resulted in voluminous documentation for each region, and have started a process of integration that ideally allows a rigorous comparison of economic, environmental and other forest values in order to allow decisions to be made about the extent in these forests of reserved, logged or other-use areas.

The New South Wales Southern Regional Forest Forum was initiated for regional stakeholders to share information and offer recommendations in regular meetings. A multi-criteria analysis was carried out with the participation of the forum members in order to show how such an approach could help the government's systematic analysis.

PEOPLE AND BIODIVERSITY POLICIES – ISBN 978-92-64-03431-0 – © OECD 2008

Three sub-criteria were developed for the analysis:

- Conservation of environmental values: biodiversity, old growth, wilderness, water and soil resources, adequate hazard reduction, forest contribution to global carbon cycles, productive capacity, health and vitality of the forest.
- Maintenance of long-term economic benefits: productive capacity, health and vitality of the forest, timber values, minerals values, apiary values, other product values, employment and community needs, recreation and tourism.
- Maintenance of social and cultural values: employment and community needs, recreation and tourism, national estate, cultural and heritage values, and indigenous values.

Table 2.12 is the impact table developed by the forum for five forest logging volume options.

Table 2.12. **Impact table of five forest options**

Indicator	Option 1	Option 2	Option 3	Option 4	Option 5
Volume of sawlogs for the next 20 years (m^3)	32 000	35 000	45 000	55 000	65 000
Achievable target met in dedicated reserves (%)					
Forest ecosystems	80	67	62	61	60
Old growth forest	73	55	52	52	52
Fauna	79	73	72	68	66
Flora	84	74	71	67	63
Wilderness reserved (%)	90	88.5	88.1	87.7	87.2
National estate areas reserved	All	All	All	Some	Some
National estate values	High	High	High	Medium	Low
Total direct mill employment	140-145	144-151	172-180	195-200	213
Total harvest and haul employment	36	38	43	50	55
Gross value output (AUD m)	15.5-16.5	16.4-17.6	19.3-22.3	22.2-25.4	25.0-27.6
Change in other employment	–48	–33	+16	+69	+115
Total gross value output (AUD m)	3 327.6	3 329.1	3 333.8	3 338.6	3 343.2

Source: Proctor, 2000.

The MCA process revealed that the option preferred by the highest number of forum members was Option 1, followed by Option 5. This result reflects the polarised nature of the forum members' priorities and shows how the trade-off in the forest management debate is between conservation and employment. The outcome of the analysis shows that middle ground choices (Options 2, 3 and 4) may not be accepted by any stakeholders, because none of them performs strongly enough in either the conservation or the employment criteria (although implementing one of the extreme cases might cause extreme opposition from the minority).

The case illustrates the value of MCA for revealing the trade-offs between economic, biodiversity and social criteria under different scenarios. The interests and attitudes of the stakeholders will determine which scenario is supported. In this case supporting nature conservation results in lower employment in the region, so social/economic criteria are in conflict with nature management criteria.

Example 2: Multi-criteria analysis with stakeholder involvement in the Buccoo Reef Marine Park, Tobago (Brown et al., 2001)

MCA is appropriate for multiple use, complex systems, such as marine protected areas, where many different users are apparently in conflict and where there are linkages and feedbacks between aspects of the ecosystem and economy. Researchers used multi-criteria analysis in the Buccoo Reef Marine Park in Tobago and involved stakeholders at all stages (they call their method "trade-off analysis"). Stakeholder analysis was carried out and social, economic and ecological criteria were identified. The impacts of four development scenarios were evaluated for these criteria, and weights were assigned by stakeholders. The scenarios were the following: (A) limited tourism development without complementary environmental management; (B) limited tourism development with complementary environmental management; (C) expansive tourism development without complementary environmental management; and (D) expansive tourism with complementary environmental management. Table 2.13 shows the estimated impact matrix of the four scenarios.

Table 2.13. **Estimated impact matrix**

Criteria	Scenario			
	A	B	C	D
Economic				
(1) Economic revenues to Tobago (USD million)	9	11	17	19
(2) Visitor enjoyment of the park (USD million)	1.2	2.5	0.9	1.7
Social				
(3) Local employment (number of jobs)	2 500	2 600	6 400	6 500
(4) Informal sector benefits (score)	5	4	3	2
(5) Local access (score)	6	5	6	7
Ecological				
(6) Water quality (μg N 1^{-1})	1.5	1.4	2.2	1.9
(7) Sea grass health (g dry weight per m^2)	18	19	12	15
(8) Coral reef viability (% live stony coral)	19	20	17	18
(9) Mangrove health (ha)	65	73	41	65

Source: Brown *et al.*, 2001.

PEOPLE AND BIODIVERSITY POLICIES – ISBN 978-92-64-03431-0 – © OECD 2008

The results revealed a consensus around the limited tourism development options for the area surrounding the park, in association with the implementation of complementary environmental management (option B). Table 2.14 shows the stakeholder weighting for the different criteria. It indicates that social issues were weighted higher than economic growth criteria, but they received lower grades than ecosystem health criteria.

Table 2.14. **Stakeholder weighting for different criteria**

Stakeholders	Economic growth	Social issues	Ecosystem health
Bon-accord Village Council	22	32	47
Buccoo Village Council	25	35	40
Department of the Tobago House of Assembly	19	29	52
Fishers	18	40	43
Recreational users	9	32	59
Reef tour operators	27	32	42
Water sports/dive operators	23	15	63

Source: Brown *et al.*, 2001.

This analysis highlights that trade-offs usually appear in different scenarios between economic, social and ecological criteria. In this case, distributive issues can be identified among both the economic and the social aspects. If ecosystem health has the highest priority, as in this case, then economic revenues and employment are lower in the most preferred scenario. Involving stakeholders can make this outcome acceptable for local stakeholders (all stakeholders also weighted ecosystem health higher than social issues and economic growth).

Example 3: Using multiple criteria to aid decisions in Šúr Wetland Nature Reserve, Slovakia (Chobotova and Klunkova-Oravska, 2006)

The Šúr Nature Reserve was established in 1952 and is one of the oldest protected areas in the Slovak Republic. It is situated in the south-west of Slovakia between the Small Carpathian Mountains and the Podunajska lowland, and is 20 km from Bratislava, the capital. The reserve is an alder forest with marshes and swamps. To gather information for the collective development and implementation of a sustainable management plan for the Šúr nature reserve, a multi-criteria decision aid analysis was undertaken in the area, involving experts and stakeholders. The main stakeholders were representatives from the national administration of the nature reserve, the district office of the nature protection authority, the local municipality, NGOs, the research station and holiday home owners. The following steps were taken: 1) development of hypothetical options for future development of the nature reserve; 2) stakeholders' analysis (conflict analysis) to investigate the structure of power and interests; 3) multi-criteria experts' analyses of main conflicts and problems; and 4) comparison of both experts' and stakeholders' views.

The following options were ranked by the stakeholders:

- A0: Non-action option, current uses would continue without any change in decision-making, management, or nature conservation practices, and there would be weak communication between stakeholders.
- A1: Integrated management of the reserve, with collective decision-making (municipality and national administration of nature reserve), a focus on nature conservation and sustainable tourism for economic development.
- A2: Integrated management of the reserve, with the national nature reserve administration acting as a legal body (after buying land from individual owners, all competencies in nature conservation and economic development would shift to the national administration).
- A3: Strict conservation-oriented option, concentration on research and education only.
- A4: Abolition of nature reserve, free economic development and elimination of nature conservation, which would exclude the reserve from international programmes and networks of protected areas, *e.g.* the Ramsar List.

Respondents evaluated each option using a scale of one to five (5 = very good; 4 = good; 3 = moderate; 2 = bad; 1 = very bad).

Option A2 (integrated management of the nature reserve with state governance) received the highest score (27% of the votes), followed by Option A1 (shift of competencies to collective decision making; 26%). The conservation option (A3) received 21% and no action (A0) only 17%. Abolition of the nature reserve and unlimited economic development of the protected area (A4) received 9%. However, positions differed substantially among the various stakeholders. Integrated management of the park with collective decision-making (A1) had positive values for all stakeholders except the district office, administration of the nature reserve and the research station. The conservative attitude of these three stakeholders (all of whom gave high priority to Option A3) is paralysing reform of the management of the nature reserve. However, present management (Option A0) is understood to be inadequate by all stakeholders and experts.

The application of MCA was key for understanding the need for a policy decision-making process that would bridge the two major opposing coalitions: conservationists who reject integrated management and supporters of integrated management (co-operation group) with both collective and state leadership.

2.5.3. Social impact assessment and stakeholder analysis

Any biodiversity-related policy can be assessed using social criteria to show how the proposed policy will affect different economic sectors or social groups. Social impact assessment (SIA) is an "umbrella" method for all the previously introduced methods.

The social impacts of a proposed biodiversity policy can be assessed by economic sectors. The worst affected sectors might be agriculture, fishing, hunting, water management, tourism, transport, energy, mining, oil, chemicals, pharmaceuticals and tourism. The income and profit level of the sectors might change if their activities are limited in time or in scale, or if they are required to pay for any environmental damage that they cause. Only a complete accounting of all impacts can ensure that society makes an informed trade-off.

Effects on local communities and the general public can be measured in many dimensions, for example:

- By regions (*e.g.* the proposed regulation will have different impacts on regions with higher natural values).
- By income categories (*e.g.* access fees to national parks can have different impacts on families in different income categories).
- By occupational categories (*e.g.* where industries are restricted and their move to other areas may cause employment problems).

The impact assessment can be accompanied by a stakeholder matrix depicting the main stakeholders of a proposed policy, the main interests of the groups, their potential influence on the process, their relationship with other groups and the possible attitudes in negotiation. A fictional matrix is shown in Table 2.15.

Table 2.15. **Stakeholder assessment matrix**

Stakeholder groups	Characteristics (potential influence on/relations with other groups)	Main interests (promote/oppose)	Main influence of the policy on the group	Economic costs and benefits of the policy
Stakeholder 1				
Stakeholder 2				
Stakeholder 3				
Stakeholder 4				

Example: Interests of stakeholders involved in conservation activities in Royal Bardia National Park, Nepal (Brown, 1998)

Royal Bardia National Park covers 968 km^2 in the mid-western region of Nepal. The area consists of *sal* forest and areas of grassland and provides habitat for a number of globally endangered species, including the Bengal tiger and the Asian one-horned rhinoceros. Since the 1990s land use and cultivation have been intensified and conflicts between conservation and agricultural production have increased. This case study identified different stakeholders and interest groups involved in biodiversity conservation and analysed the policy prescriptions they are likely to promote. Table 2.16 shows the main characteristics of the local stakeholder groups.

Table 2.16. **Stakeholder matrix, Royal Bardia National Park**

Group	Scale of influence	Source of power	Interests/aim	Means
Indigenous people	Local	Very limited	Livelihood maintenance, use protected areas for subsistence needs; minor trading of products: thatch, fodder, building materials, fuel, wild foods, plant medicines; hunting and fishing	Subsistence farming, minor marketing, legal and illegal extraction of resources from protected areas
Migrant farmers	Local	Limited	Livelihood maintenance, use protected areas for subsistence needs, thatch, fodder, fuel, building material	Cash farming plus subsistence, legal and illegal extraction of products from protected areas
Local entrepreneurs	Local	Many hold official positions locally	Profit, commercial, range of small enterprises, tourist and non-tourist based	Small business enterprises, buying and selling to tourists
Tourist concessions	National/some international	Lobbying/may hold official positions	Profit, commercial, expansion, some revenue may be earned overseas, control tourists staying in protected areas overnight	Tourist revenues, concessions from government
Government conservation agencies	National	Administrative and supervisory	Conserving wildlife and facilitating tourist development	Enforcing park boundaries, imposing fines
Conservation pressure groups	Local, national, some international links	Lobbying, may have personal contacts, international funding	Conserving biodiversity but with consideration for local livelihoods	Lobbying, publicity
International conservation groups	International	International funding, green conditionality	Conserving biodiversity, limited interest in human welfare	International legislation, lobbying

Source: Modified from Brown, 1998

The analysis shows that biodiversity-related policy measures are likely to have different implications for different stakeholder groups. While many local stakeholders use biodiversity for subsistence and for business, there are likely to be negative distributive effects of conservation policies that government needs to deal with. The stakeholder matrix helps identify the main groups, their interests, means and power.

2.6. Summary and comparison

Table 2.17 compares the most important methods for measuring the distributive effects of biodiversity policies, and describes when they are most appropriate.

Table 2.17. **Advantages and disadvantages of the key methods for measuring distributive effects of biodiversity policies**

Method	Strengths	Weaknesses	Applications	Examples
Measures for equality of income distribution (Lorenz curve, Gini coefficient)	Graphical illustration and numerical measure of equality	Cannot be used in a very complex situation. Large statistical data requirements	Can be used for a well-defined group	Measures of equality of income distribution impacts of privatisation of mangroves in Vietnam
Extended CBA (by stakeholder groups)	Gives quantitative results segregated by stakeholder groups	Extensive statistical data are needed	At both national and local level, where income or stakeholder groups can be easily identified. Where monetary valuation of both costs and benefits is possible	Three potential park scenarios in the Ream National Park, Cambodia
Social accounting matrix (SAM)	Shows the flow of income from one sector to another	Extensive statistical data are needed; rather complicated method. Problematic where no financial information is available	Can be used in local, regional and national circumstances	Distributional impacts of alternative forest management of the Upper Great Lakes region, USA
Distributional weights	Can compare efficiency and distributive impacts on a common scale	Extensive statistical data and assumptions about utility function are needed	Both national and local level where income groups or stakeholder groups can be easily identified. Where monetary valuation of both costs and benefits are possible	*UK Green Book*
Atkinson inequality index	Uses normative judgements about social welfare	Has been used only in the narrow field of income studies. Applicability to biodiversity policies is still an open question	Can be used at international, national, and local levels to the extent that normative judgements can be plausibly applied in the chosen context	No example related to biodiversity policy yet

Table 2.17. **Advantages and disadvantages of the key methods for measuring distributive effects of biodiversity policies** *(cont.)*

Method	Strengths	Weaknesses	Applications	Examples
Employment-based analysis	Unconventional, straightforward measure of the level of employment	Income changes cannot be measured by this method. Cannot capture other social effects. May hide important dimensions of job status, qualification match and labour market frictions	In rural areas, where employment changes are more important than the changes in the income level	Measuring the employment impact of different farming activities Yucatan, Mexico
Child nutritional health status	Unconventional, straightforward measure of the nutritional status of children	Nutritional status depends on many factors. Former calculations had statistical and other problems	In developing countries, where it is difficult to use other measures and where nutrition can have direct link to biodiversity policy measures	Measuring nutritional status of children in marine protected areas in the Philippines
Stochastic dominance analysis	Multidimensional analysis of the distribution of social welfare (not only income and wealth, but others, *e.g.* education or health)	Strong assumptions are needed about how the dimensions relate to welfare, and social weights need to be given to different dimensions	At local, regional or national level	The Human Development Index of the United Nations
Multi-criteria analysis (MCA)	A wide range of distributive effects can be measured using social and economic criteria: the level of measurement is not a problem. It can be a base for further discussion with stakeholders and for assessing trade-offs	Results heavily depend on the weights given to the criteria (weights can be given by experts, stakeholders or policy-makers)	In local, regional and national situations, and in complex cases when many criteria need to be taken into account, and where some of the effects cannot be measured in monetary terms	Australian forest policy. Buccoo Reef Marine Park, Tobago. Šúr Wetland Nature Reserve, Slovakia
Social impact assessment (SIA) and stakeholder analysis	Impacts on stakeholders and distributive effects are assessed. All other methods can be used for the assessment	Sometimes it is superficial and lacks monetary data (if there is no clear guidance or indicators given)	In any policy situations at local, regional and national level	Stakeholder analysis in the Royal Bardia National Park, Nepal

The table shows that the applicability of the method depends on the policy measure, the geographical scale and data availability. This is a key message emerging from the literature on assessing distributive impacts: the method chosen cannot be separated from the policy to be analysed.

If policy evaluation budgets are well funded, a mix of different methods can be used. On the one hand, this gives policy-makers strong empirical foundations on which to make their policy choice. The information gathering required for broader and more encompassing forms of welfare assessment is also highly compatible with the types of consultative and participatory approaches advocated in Chapter 6. On the other hand, the literature on assessing distributive impacts also reveals that measures of inequality are not unequivocal: while a project may decrease inequality as measured by one method, the very same project may increase inequality according to a different method of measurement. Different measures have therefore to be given different weights, whether implicitly or explicitly, and while these weights should reflect society's preferences concerning trade-offs between different dimensions of welfare (Trannoy, 2003), they will naturally contain elements of both arbitrariness and dispute.

This chapter has looked at a range of methods for assessing the distributional impacts of biodiversity policies. In the next chapter we look at some of these impacts, discussing them as static issues, reflecting a snapshot in time of the impacts. In Chapter 4 we then discuss them as intertemporal issues, reflecting the impact across generations.

Notes

1. Welfare economics is an approach to assessing the desirability of different outcomes grounded in utilitarian philosophy (Hanley and Spash, 1993).
2. A simple example is when government needs to raise a fixed amount of revenue. Suppose there are two groups who can be taxed to raise the needed revenue and who are similar in terms of income. If the government taxed only one group, it is generally the case that their loss of welfare will be greater than the combined loss of taxing both groups at a lower rate.
3. Quintiles are those points 1/5, 2/5, 3/5 and 4/5 of the way through a frequency distribution.

ISBN 978-92-64-03431-0
People and Biodiversity Policies
Impacts, Issues and Strategies for Policy Action

Chapter 3

The Distributive Effects of Biodiversity Policies: Static Analysis

3.1. Biodiversity policies: process and instruments

At an abstract level, biodiversity policies are about change. Successful policies change the way society interacts with the natural environment so that society gains most from the use and/or conservation of biologically diverse habitats and ecosystems. These potential welfare gains from changes in the use of habitats and ecosystems are the foundation for, and deliver political legitimacy to, specific biodiversity-related policy objectives. The starting point for such biodiversity policies is the realisation by policy-makers that such welfare gains can be achieved. Policy objectives then must state the outcomes to be accomplished. Typical examples of such objectives are "the preservation of a genetically viable population of *Maculinea* in its natural habitat up to the year 2100" or "a reduction of pesticide exposure of songbirds in the UK by 10% by 2010".

Policy objectives only specify outcomes; they do not tell the policy-maker how to achieve them. Once the objectives are defined, therefore, the next step for the policy-maker is to choose a suitable set of instruments for enhancing habitats and ecosystems at the aggregate level. For biodiversity policies, these instruments cover a wide variety of possibilities which can be grouped into three classes (see Section 3.2.2 for a more detailed list):

- Instruments targeting the behaviour of individuals, households and groups interacting with habitats and ecosystems through – for instance – extractive activities such as gathering firewood. The nature and volume of these interactions are assumed to be determined by their relative costs and benefits. Instruments that change these costs and benefits will therefore have an indirect effect on what activities are carried out, since individuals and households will find it in their interest to adjust their activities in response to these changes. Typical examples of instruments that target behaviour are changes in rewards for certain activities, *e.g.* the removal of subsidies or the payment of conservation premiums; an increase in cost of carrying out certain activities, *e.g.* through the imposition of a tax on harmful activities; the provision of information to individuals and groups, *e.g.* through market research; or the creation of outside institutions that increase the value of engaging in certain practices, *e.g.* eco-labels on products that have been produced in compliance with certain environmental standards.

 PEOPLE AND BIODIVERSITY POLICIES – ISBN 978-92-64-03431-0 – © OECD 2008

- Instruments targeting the institutions that govern the constraints imposed on and the opportunities available to individuals. Institutions are "the humanly devised constraints that shape human interaction" (North, 1990) and have emerged over the last ten years as one of the truly significant factors determining the efficacy of conservation policies (Barrett *et al.*, 2005). Typical examples of institutional changes are alterations in property rights assignments, *e.g.* by restricting the right to let animals graze on land, or devolution of management rights from government agencies down to communities.
- Instruments that sever the links between human populations and ecosystems or habitats. One example is the removal of individuals, households and groups from the habitats and ecosystems that are to be protected through park eviction or resettlement policies.

The choice of the appropriate instruments for protecting biodiversity has been widely discussed in the literature (see OECD, 2004; van Kooten and Bulte, 2000 for surveys). The common criterion for choosing one instrument over another is that of efficiency: the instrument should be chosen that protects biodiversity at minimum aggregate cost to society. Since instruments differ in their aggregate cost depending on the policy context, this criterion has important practical repercussions for policy-making, ensuring that policies are not implemented in a wasteful fashion.

The focus in this book on evaluating instruments is somewhat different. What is less relevant here are variations in aggregate costs between instruments; instead we are concerned with how each instrument affects the welfare of those affected by the policy. Instrument choice will therefore be considered not only with reference to aggregate costs, but also with reference to those on whom the final costs will fall.

Figure 3.1 summarises the policy-making steps in the form of a "linear policy-making model", which includes the links to welfare impacts. The key relationship is between the grey box on the right and the potential aggregate welfare improvements on the left of the diagram. The grey box describes how the interaction of habitats and ecosystems with people under a given set of institutions can generate social well-being. Policy-making is then about realising that by changing elements of the interactions within the grey box, society could be lifted to a higher level of well-being. An example would be the creation of a protected area that will preserve additional habitats. This gain in society's overall well-being is the potential aggregate welfare improvement in the left-hand box. The aggregate improvement, however, consists of many individual welfare impacts. Gains to some households from increasing biodiversity coexist with losses to others who bear the costs. Aggregating across all the individuals, households and groups commonly loses this fine detail of losers and winners. Instead, policy-making is generally advocated whenever the potential **aggregate** welfare improvements are large and positive. These welfare improvements are then translated into a policy objective to be implemented at least cost using the appropriate set of instruments.

Figure 3.1. **The linear policy-making model**

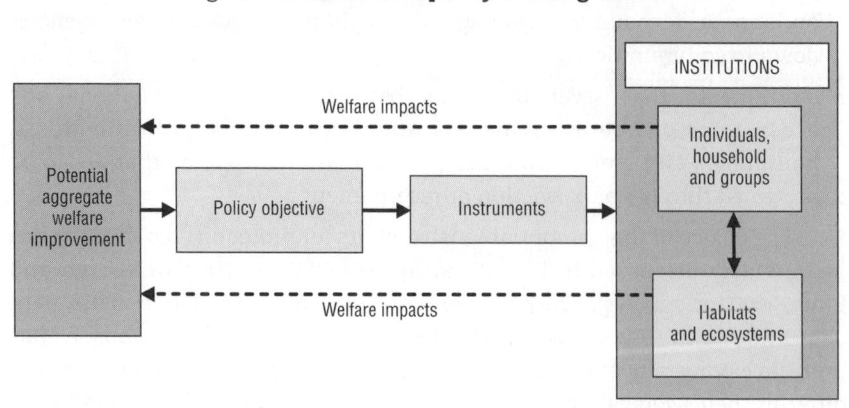

Apart from a summary of the step-wise procedure of the linear process, the linear policy-making model contains a few additional messages about the relationship between policy objectives, biodiversity policy outcomes and distributive outcomes that have been emphasised repeatedly in the literature (see Barrett *et al.*, 2005 for a survey):

- The link from objectives to outcomes is not direct, but modulated by various intervening factors inside the grey box.
- The grey box comprises context-specific factors, such as the individual preferences of local residents, local land use conditions and existing institutions that constrain and condition individual behaviour, such as the distribution of property rights and presence of monitoring.
- These specific factors imply that understanding the environmental and distributive impacts of certain policy measures requires detailed information about the functional linkages between these factors.
- Instrument choice is critical in determining policy outcomes, and the interaction between instrument choice and institutional factors needs to be closely studied.
- The welfare impacts of biodiversity policies tend to be very different for those individuals interacting directly with habitats and ecosystems than for the public at large.

 PEOPLE AND BIODIVERSITY POLICIES – ISBN 978-92-64-03431-0 – © OECD 2008

- Policy objectives determine the primary effects of a policy, which depend on the size of the increase in biodiversity-related goods and services and the size of the costs of the project. As a rule, the more different the project is to the situation before, the greater the primary effects.
- Primary effects can be neutralised, mitigated or amplified by secondary impacts. Secondary impacts are a function of the instruments used to carry out the policy objectives. As a rule, the more coercive and less reward-based the instrument, the more accentuated the distributive effects of the policy (see Section 3.2.2, "Secondary policy effects: the role of instruments").

3.2. The distribution of biodiversity net benefits

Here we examine first the primary and then the secondary distributive impacts of biodiversity policies on the relative well-being of those affected. The primary welfare effects are simply a function of the change in the supply of biodiversity-generated goods and services brought about by the biodiversity policy and the cost of providing this change in supply. The extent to which these changes enhance or reduce welfare of (groups of) individuals is determined by the relationship between the demand for biodiversity-generated goods and services and levels of income and wealth. This demand differs considerably with the type of good or service that individuals consume. Even within certain classes of goods and services, attention will have to be paid to the functional relationships between habitats/ecosystems and welfare. These relationships form the core of the grey box in Figure 3.1.

3.2.1. *Primary policy effects*

The economic values of biodiversity

Biodiversity interacts with the economic welfare of individuals and households in complex ways. Biodiversity is important for purely consumptive purposes, *e.g.* in the form of ecosystems services, but also as a key production asset. From an economic perspective, the aggregate contribution of biodiversity to human well-being is summarised in the concept of total economic value (Table 3.1; Pearce and Moran, 1994).

The TEV is made up of use values (UV) and non-use values (NUV). Use values can be further divided up into:

- *Direct use values (DUV)*: goods and services that can be directly consumed, *e.g.* supply of biodiversity-related goods, recreational areas, medicinal plants, and rare megafauna. Within these categories it is worth distinguishing commercial use and harvesting for own consumption when considering distributive issues. This is because the ability to internalise negative repercussions from direct use of biodiversity-related goods and services will be stronger for commercial uses and thus less in need of corrective policy.

- *Indirect use values (IUV):* the life-sustaining and other important ecological services that are provided by biodiversity to individuals, to communities and to mankind as a whole, *e.g.* carbon sequestration, watershed management and flood protection. Indirect benefits also include the economic opportunities created by biodiversity, *e.g.* ecotourism revenues, enhanced scope for organic agriculture, and bio-prospecting. Some of these benefits cannot be measured or captured easily (only some of the resulting revenues can be counted); nevertheless they are very important.
- *Option values (OV):* possible future direct or indirect use values of biodiversity, which are currently not known. Since the number of species that exist could be two orders of magnitude higher than is currently catalogued, the potential for unknown benefits of biodiversity may be large.

Non-use values are made up of:

- *Existence values (EV):* derived from the existence value of nature, irrespective of its use. Among others, this can include aesthetic values and the moral value of passing on unspoilt nature to future generations.
- *Bequest values (BQ):* inherent in enabling people to pass habitats and ecosystems on to subsequent generations.

Table 3.1. **Categories of economic value attributed to environmental assets**

Total economic value				
Use values		Non-use values		
Direct use	*Indirect use*	*Option values*	*Bequest values*	*Existence values*
Outputs directly consumable	Functional benefits	Future direct and indirect values	Use and non-use value of environmental legacy	Value from knowledge of continued existence
Food, biomass, recreation, health	Flood control, storm protection, nutrient cycles	Biodiversity, conserved habitats	Habitats, prevention of irreversible change	Habitats, species, genetic, ecosystem

Source: Pearce and Moran (1994).

The TEV demonstrates the wide range of benefits that biodiversity conservation policies can generate for people, ranging from tangible consumption benefits such as provision of timber and food to immaterial benefits such as existence values based on the knowledge that a species is being conserved. Overall, these benefits are significant in scale and are the basis for policies that try to maintain and enhance habitats and ecosystems (Pearce and Moran, 1994). However, it is also important to understand that the TEV is a very specific way of representing the relationship between societies and their natural environment.

PEOPLE AND BIODIVERSITY POLICIES – ISBN 978-92-64-03431-0 – © OECD 2008

The TEV allows us – at the highest level of abstraction – to summarise the total contribution of biodiversity to human well-being by adding together the individual value components (Pearce and Moran, 1994). In simple terms, this means that the TEV can be represented as:

TEV = UV + NUV

= (DUV + IUV + OV) + (EV + BV)

Put this way, the TEV is a statement of the value of the biodiversity assets available on a global basis. It therefore provides a statement about their gross value.

The economic net values of biodiversity

There are two important problems with the TEV that are relevant for distributive analysis:

- It is a gross value concept: it does not include the costs, monetary and otherwise, of procuring these goods and services by managing natural resources in a specific way.
- It may be an empirically empty concept: estimating the TEV with any degree of confidence poses some considerable difficulties (see Costanza *et al.*, 1997 and the related discussion). Empirically more relevant and technically implementable is the concept of the marginal economic value (MEV). This is a measure of the change in the TEV of a habitat or ecosystem brought about by an intervention. Economic theory offers sophisticated methodologies for estimating this value in monetary terms for many of the value categories (OECD, 2002).

The analysis of distributive issues in the following sections thus builds on the ideas inherent in the TEV, but differs in three ways:

1. It is concerned not with the total economic value of biodiversity, but with the marginal economic value of a biodiversity-related policy or project.
2. It is concerned with net, rather than gross, values of biodiversity.
3. It is not concerned with an aggregate value, but with how the aggregate value is distributed.

In short, its focus is on the *distribution of the net marginal economic values of biodiversity policies*. Questions that arise in this context are: how is the (gross) MEV distributed across different groups at different levels of income and wealth? And how much do these groups sacrifice for the purpose of generating this MEV? In other words, the key concern is how the MEV of biodiversity policies is shared across different parties *net of each party's cost of maintaining the habitats and ecosystems* that give rise to biodiversity's values.

Income, wealth and the distribution of benefits

The starting point for analysing the distribution of net benefits from biodiversity policies is to recognise that households with different incomes rely on very different goods and services generated by biologically diverse ecosystems and habitats. The TEV concept does not explicitly distinguish between the private and public goods components of the biodiversity-produced goods and services. However, the distributive dimension is already embedded for the simple reason that the functional relationship between households and habitats and ecosystems differs along the income scale.

The most important interaction between low income households and their natural environment is through extractive and consumptive activities. Richer households are more likely to be interested in the public goods aspects of biodiversity as the household income base is less likely to depend directly on primary resources. Both private and public goods are therefore jointly produced in an integrated and complex way, but different income groups have very different economic perspectives on which of these outputs are more valuable to them at the margin.

Distribution of gross benefits from biodiversity policies

Biodiversity policies are motivated by the externalities inherent in the management of biodiverse ecosystems and habitats. Assessing the impact of these policies on individual well-being is a challenging task for two reasons: *i)* public goods and services related to biodiversity are not traded on markets either for consumption or as production inputs. The functional dependence of individuals and groups on biodiversity-produced services is therefore frequently not easily evident prior to the policy intervention; and *ii)* the value of these goods and services relative to other goods (the relative price) cannot be observed directly. Together this means that the contribution of biodiversity to welfare can only be fully assessed through procedures that impute the quantity and price of biodiversity-related goods and services indirectly. Here, the assessment builds on a number of empirical studies that provide guidance on the direction and volume of welfare changes brought about by biodiversity policies.

The demand for goods and services generated by biodiverse ecosystems and habitats exhibits income effects. While these effects can be negative,[1] most of the time they are positive, meaning that as incomes go up, so does the demand for those goods and services provided by biodiversity. Goods for which the income effects are positive are called "normal" goods. Thus, biodiversity goods and services are normal goods.

Income effects for biodiversity policies are the income elasticity of willingness to pay for conservation, which is the percentage change in willingness to pay for a 1% change in income. Positive elasticities imply that the rich benefit more from environmental improvements than the poor. Income elasticities of more than one indicate that the environmental good is a "luxury" good, as the willingness to pay for the good increases faster than the growth in income. For luxury goods, the distributive effects of public policies designed to increase their supply are strongly progressive, with the rich benefiting disproportionately.

Theoretically, there are three reasons for predicting a positive and significant income elasticity of the willingness to pay for biodiversity:

- Most environmental goods and services have all of the properties of "normal" goods (Baumol and Oates, 1988). Thus as incomes increase, so does the demand for those goods and services provided by biodiversity.
- Rising incomes do not only lead to a higher demand for each normal good, they also lead to a demand for more goods (Dixit and Stiglitz, 1977; Theil and Finke, 1983). The inherent variety of biodiversity-related goods and services should therefore elicit a higher willingness to pay (Bellon and Taylor, 1993).
- Increasing scarcity of rare environmental resources such as biodiversity may induce a change in preferences towards a higher marginal valuation (Krutilla, 1967).

On this basis, we would expect an income elasticity of WTP for biodiversity close to one or above. Empirical estimates for environmental goods in general and biodiversity related goods and services in particular are not plentiful. There are also good theoretical and methodological reasons for using these estimates with strong caution (Flores and Carson, 1997).

Table 3.2 summarises the results of studies and meta-studies of income elasticity of WTP for biodiversity. Without going into the details of study methodologies and econometric considerations, it is clear that there is a wide variety of estimates, ranging between values of 0.2 up to 2.

According to these estimates, the mean income elasticity of willingness to pay for biodiversity policies lies somewhere in the region of 0.5. It is therefore positive, and it is possible that it is above 1 for a number of important cases. Note that Schläpfer *et al.* (2004) find that biodiversity systematically exhibits a lower income elasticity than other environmental goods. Benefits are therefore likely to be less pro-rich than those for other environmental amenities studied empirically.

The general thrust of these findings is supported by research examining optimal public spending on park projects. Kalter and Stevens (1971) found distributive benefits favouring high income rather than low or medium income households. In Maine, Reiling *et al.* (1992) found that a combination of

Table 3.2. **Empirical measures of the income elasticity of marginal WTP for biodiversity and related projects**

Authors	Method	Elasticity estimate
Kriström and Riera (1996)	Meta-analysis of 6 contingent valuation (CV) studies	0.2 and 0.3
Schläpfer and Hanley (2003)	Referendum data on landscape amenities management	> 1
Schläpfer, F., Roschewitz, A. and Hanley N. (2004)	Referendum-format CV study on the value of landscape amenities protection	0.35
Horowitz and McConnell (2003)	Meta analysis of 23 CV studies	0.1-0.4
Hökby and Söderqvist (2003)	Meta analysis of 21 elasticity estimates from CV studies in Sweden	< 1
Borcherding and Deacon (1972)	Referenda on parks and recreation	> 1
Bergstrom and Goodman (1973)	Referenda on local public goods	> 1
Eskeland and Kong (1998)	Different environmental improvements	0.1 to 2.

providing a state park and charging an access fee has strongly discriminatory effects on low-income households. In a comparison of national and urban parks in Israel, Feinerman *et al.* (2004) found that policies favouring national parks generate disproportionate benefits for high and highest-income households. A reallocation of these funds towards local parks would be preferred by all but the richest 10% of the population. The reason is that while national parks generally provide greater conservation and recreation benefits to their users than urban or local parks, the cost of access, in particular travel costs, reduces net benefits considerably. A study in California (Kahn and Matsusaka, 1997) supports this finding, pointing out that rich households can provide private substitutes for urban parks, thus explaining the negative income elasticity of urban parks for the high-income sections of society.

These findings suggest there are strong theoretical and empirical reasons for predicting that the primary (or "first-round") benefits of biodiversity policies will accrue to a greater extent to households with higher incomes.

This line of discussion, however, should not be over-interpreted. Rapid growth rates in many developing countries (*e.g.* China and India) suggest that they too will soon have higher demand for non-use biodiversity-related benefits. As will be discussed below in the inter-temporal analysis, when policy contexts are changing over time, policies that appear not to be pro-poor today may indeed benefit poorer individuals in a predictable way in the future. Thus, the picture drawn here is of a static distribution of benefits.

Distribution of costs from biodiversity policies

Dixon and Sherman (1991) categorise the costs of biodiversity policies as follows:

- Direct costs: the cost of implementing the policy. These are usually paid for out of public funds which have to be raised by general taxation or, where possible, out of payments such as user (or entrance) fees. Examples include costs of managing the protected area (salaries, cars), demarcating a conservation area, and monitoring and enforcement.
- Indirect costs: non-budgetary, material costs arising from the implementation of the policy. Typical examples are crop losses at the boundaries of protected areas as a result of increased wildlife population levels within.
- Opportunity costs: the value of alternative uses foregone by virtue of implementing the policy. Opportunity costs arise through the immediate sacrifice of consumption possibilities previously exercised and no longer possible; and through the delayed sacrifice of potential future gains that would have arisen from alternative uses of the existing assets. This second component, however, varies strongly with the degree of irreversibility of the policy.

Direct costs. Direct costs of biodiversity policies fall on governments, in the case of public projects, or on dedicated funds and their sponsors, in the case of non-governmental projects. Establishment and management costs are key components. Balmford *et al.* (2000) provide estimates for the establishment of conservation areas on an international basis. These costs are generally estimated as small relative to other costs and benefits involved in biodiversity policies.

The most important determinant of the direct cost of biodiversity policies is instrument choice. Governments have control over the amount of public funds required for projects by varying the instrument used. If biodiversity policies involve the actual acquisition of land for conservation, then these costs will be much higher than outlays for establishment and management (Balmford *et al.*, 2003). Some instruments, such as outright takings (where constitutionally permitted), involve little or no direct cost to the government. For other instruments, such as user fees, money will flow through the government's hands, but the government's direct costs are purely administrative.

When governments incur direct costs and fund them out of general taxation, their distribution within the country's population is the same as the distributive effects of raising public funds (Kriström, 2006). The distributive incidence of direct costs at a national level can therefore quickly be determined by understanding the properties of the general taxation system (Kriström, 2006).

Indirect costs. Indirect costs, though likely to be smaller than opportunity costs overall, follow a similar logic. Exposure to conservation-induced damages will be higher for those more reliant on extractive and consumptive activities in, or adjacent to, the conservation area.

An example comes from protected areas in the Nanda Devi Biosphere Reserve. Maikhuri *et al.* (2000) report mean annual agricultural losses due to wildlife of USD 90.6 per household for areas adjacent to the protected area, compared to USD 27.9 for areas further away. Villages closer to the protected area therefore bear greater costs. Along the income dimension, villages did not significantly differ from each other in terms of income averages. Within villages, however, there is evidence of regressive effects since poorer households in the villages were more dependent on agricultural activities than richer households (Maikhuri *et al.*, 2000). A similar theoretical analysis of the Serengeti National Park was done by Johannesen and Skonhoft (2004).

Opportunity costs. Opportunity costs are considered a dominant component in the distributional incidence of biodiversity policies (Dixon and Sherman, 1991) and have an explicit distributive dimension. Theoretically, we would expect that with strict conservation policies focusing on the broader public goods of habitats and ecosystems, the willingness to pay to avoid the implementation of conservation policies is higher for poorer individuals and households. This is the case for two reasons:

- Poorer households are more likely to rely to a greater extent on the extractive and consumptive components of ecosystems and habitats as they are more likely to be active in the primary sector (Naidoo and Adamowicz, 2006b).
- Poorer households have a smaller asset base than richer households (comprising physical and human capital) and are more likely to suffer from missing markets; as a result low income households have fewer alternative income generating options (Reardon and Vosti, 1995).

This theoretical argument is supported by a number of empirical studies. Ferraro (2002) studied the local costs of the establishment of the Ranomafana National Park in Madagascar in 1991. Using household-level data on agricultural and forest use, along with qualitative data, his empirical research demonstrated that even when conservatively estimated, the opportunity costs of conservation for residents living adjacent to the national park were substantial. The aggregate net present value of opportunity costs was USD 3.37 million, which translates to household level NPV costs of between

USD 350 to USD 1 300, or an annual loss of between USD 19 and USD 70. These costs are substantial, at between 1.5 and 6% of annual income.

Reddy and Chakravarty (1999) studied income and forest dependence in Northern India in the context of the possible introduction of forest conservation measures restricting consumptive and extractive activities. By disaggregating income sources, the authors demonstrated that the income share of forest-related consumptive and extractive activities is strongly negatively related to income (see Table 3.3).

Table 3.3. **Poverty indices, with and without forestry**

H[1]	I	S	P^C	P^C	FGT
			ε = 0.25[2]	ε = 0.25	(± = 2)
1) Poverty indices with forestry income					
0.432	0.345	0.199	0.157	0.176	0.097
2) Poverty indices without forestry income					
0.528	0.426	0.295	0.239	0.271	0.124

1. H – Head count ratio, I – Income gap ratio, S- Sen index, P^C -Clark, Heming and Chu ration, FGT – Foster, Greer and Thorbecke index.
2. The symbol ε stands for a poverty aversion parameter.

Source: Reddy and Chakravarty (1999).

This study in Northern India raises the possibility that biodiversity policies that entail restrictions on forest use can have significantly regressive impacts and thus lead to an increase in inequality and poverty.

Naidoo and Adamowicz (2006b) assessed the opportunity costs of conservation policies in terms of agricultural production and timber extraction foregone for forest areas in Paraguay. Their study area was characterised by smallholders on land with lower productive value, but who were more dependent on agricultural income than larger operations. This raises a problem for policies: land with lower productive value should be used first for conservation, but this would affect a group of landowners less affluent and more reliant on the productive use of these areas.

While the balance of the evidence points to a general pattern of regressive incidence, recent studies have highlighted some subtleties that are of relevance in the context of distributive impacts. The most important is that there is considerable heterogeneity among households, even within a single village, in the relationship between income levels and opportunity costs of conservation policies. Alix-Garcia *et al.* (2004) stressed the influence of institutional factors on the interaction with the natural resource base adopted at the village level. In a study of communal lands (*ejidos*) in Mexico, the authors found that WTP to avoid conservation is significantly higher in communities who have invested in forestry extraction activities. Coomes *et al.* (2004) studied the empirical relationship between volume of extraction from tropical forests and its share of income at the household level. The authors found that within each community, reliance on extractive activities can be heavily concentrated among a very small number of households. These observations point to significant heterogeneities of conservation policy impacts that may not be directly related to income levels.

The primary distribution of net benefits. The previous section looked at the distribution of gross benefits and costs of biodiversity policies across various income groups. This section integrates these findings into an assessment of the primary incidence of net benefits (gross benefits minus costs) across different income groups. The theoretical arguments and empirical evidence deliver a complex picture of different patterns of incidence depending on the category of benefits and costs considered. On balance, the evidence suggests that primary effects of biodiversity policies (*i.e.* before being mitigated or amplified through different instruments) are characterised by:

- Progressive benefits, *i.e.* higher income households are likely to be willing to pay more for biodiversity policies to be implemented than lower income households.
- Regressive costs, *i.e.* lower income households are likely to be willing to pay more to avoid biodiversity policies being implemented than higher income households.

Figure 3.2 provides an example of a highly simplified and stylised graphical representation of the primary distribution of gross benefits, costs and net benefits in relation to income. The progressive benefits of biodiversity policies are depicted as the benefit curve, which increases with income, at the top of the diagram. At the bottom are the (moderately) regressive costs of the policy, which increase as income declines. The net impact for the representative household or individual at income I is then the difference between benefits and costs for that income level. Assuming a uniform density of individuals along the income interval from zero to I_m, the highest income in this society, the areas G and L are appropriate measures of the aggregate net benefit gain and loss to society. Since G > L, the policy is welfare improving at the aggregate level and hence admissible under a CBA efficiency criterion.

Overall, the combination of progressive benefits and regressive costs leads to strongly regressive net benefits, as can be seen from the fact that the net benefit curve in Figure 3.2 is steeper than the benefit curve. Specifically, it results in those households within society with an income below I' being primary net losers from the biodiversity policy. Under standard assumptions

Figure 3.2. **Example of net benefits and their distribution under progressive benefits and regressive costs**

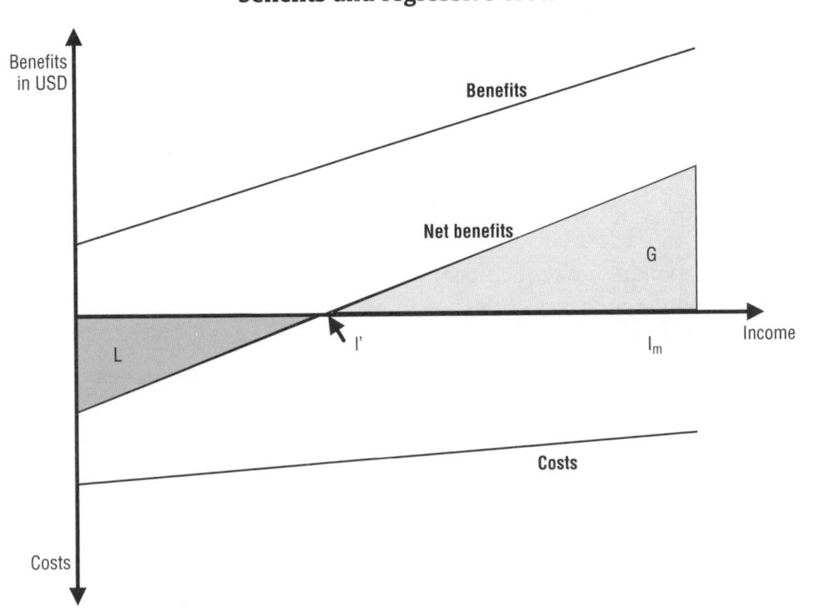

about the relationship between income and welfare, policies increase – at the primary level – welfare inequality.

These observations lead to the conclusion that "biodiversity conservation policies are probably not pro-poor on a net benefit basis" (Deacon, 2006). This means that at a first level, the distributive impacts of biodiversity policies that enhance the supply of biodiversity-related goods and services will typically generate greater benefits for those better off, frequently imposing net costs on those less well off. These primary impacts, however, do not represent the ultimate distributive outcomes. This is because biodiversity policies need to be implemented and in the process of implementation, instruments need to be chosen. As will become clear in the discussion on secondary policy effects, instrument choice is an important modifier of the primary impacts of biodiversity policies. It has the potential to produce different distributive patterns of welfare outcomes by channelling gains and losses in particular directions. This is the topic of Section 3.2.2.

Distribution across developed economies. Inequality in income and economic development is one important context for considering distributive issues. There is, however, another context in which distributive impacts occur: where biodiversity endowments vary amongst similarly developed countries. Many OECD countries, for example, have similar incomes, but different biodiversity assemblages, and still have the potential for significant distributive issues. In this section, we provide a concise explanation as to how that redistribution occurs.

Consider the case of a hypothetical world containing two countries of equal income and roughly similar preferences for biodiversity. Two simple cases arise when considering the link between the benefits of biodiversity and the amount of biodiversity that is available. Gains from biodiversity for the people in each country either stay constant (curve A in Figure 3.3) or decrease (curve B) with each additional unit of biodiversity conserved.[2] These cases allow us to explore how much benefit each country gains from the existence of biodiversity in other countries.

Figure 3.3. **Distributive issues among similar countries**

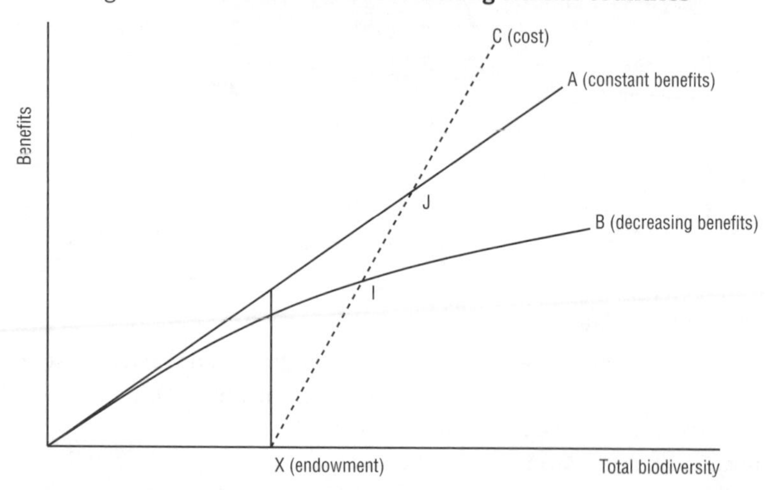

If X is the endowment (*i.e.* the state of biodiversity in some base year), then the desire for additional biodiversity will be stronger with A than with B – because A shows more benefit from biodiversity than B. For simplicity, imagine that two countries are similar in income and tastes. Adding to biodiversity requires trade-offs to be made because there are costs. Again for simplicity, we take the case where maintaining existing biodiversity is costless, but adding to it incurs costs. The dashed cost line in the figure thus starts at the endowment X and increases from there since only additions to the endowment incur costs. The costs are increasing so eventually the cost curve meets the benefit curves for both A and B. If the cost and benefits curves include all social aspects of costs and benefits, then the intersection of the cost and benefit curves is what government policy should strive to achieve: cost equals benefit. In the case of constant benefits (curve A), this would be at

J, which gives an amount of biodiversity indicated on the horizontal axis. Notice that since the benefits curve is a straight line, each addition to biodiversity brings the same gain, and a simple "adding-up" of biodiversity gives the total global benefit. When the two countries both move respectively to J, citizens in both countries gain. In that case, if each country simply supplies the "right" amount of biodiversity for its own citizens, then even though biodiversity is a public good, the right global amount of biodiversity will be supplied. There is no co-ordination problem in supplying biodiversity. By acting "locally" each country solves a "global" problem.

On the other hand, if benefits are decreasing – as with curve B – then each country's benefit from the last unit of biodiversity that it supplied will be different if endowments are different. This is because the location of the optimal point of provision I within each country will vary with endowment and the benefit from the last unit supplied will be different – the co-ordinated solution will be different from the unco-ordinated solution. As biodiversity is a global public good and endowments differ, then even when incomes and preferences are similar between countries, there is a redistributive impact from what each country does and there is a co-ordination problem. A general result for the global provision of a public good is that each country should pay according to marginal benefit until the sums of those benefits (at the margin) cover the cost of additional biodiversity (discussed further in Section 5.2.1).

3.2.2. Secondary policy effects: the role of instruments

While the primary policy effects presented above are determined by the objectives of biodiversity policies, instrument choice introduces another source of equity impacts of biodiversity policies. These are referred to as "secondary distributive effects". Here we are particularly concerned with the direction of the secondary distributive effects, *i.e.* the extent to which specific instruments mitigate or amplify the primary policy effects.

To put the impact of these instruments into their proper context, we need to understand the existing rights of those affected by the policy. We therefore start with a short digression into the types of property and use rights likely to be encountered by the biodiversity policy-maker. We then give an overview of the most important biodiversity policy instruments, and classify these instruments in preparation for an overview of theoretical and empirical findings on their distributive impacts.

Property rights and instrument choice

Many of the common policy instruments in biodiversity policies affect ownership and other use rights related to natural resources. When the policy is implemented, individuals, households and groups will value these natural assets differently depending on their and other's ownership rights to them. As a result, pre-policy rights are an important determinant of the distributive impacts of policies. There are four main types of ownership rights:

- Open access to the resource (no direct ownership): no user can be excluded from enjoying the benefits of the resource, *e.g.* fisheries in the high seas, or the former use of tropical forests in developing countries. Natural constraints exist, however, in the form of limited regenerative capacity.
- Private ownership: generally entitles the owner to exclude others from the benefits of the asset. In practice, the specific scope of private property is more tightly circumscribed and differs from country to country. For example, the purchase of real estate in some OECD countries gives owners broad rights to do more or less as they wish with the land. In other countries, however, it gives only a limited set of rights which include very tight constraints on the "owner" and which give the government greater scope to act. Alienability of an asset, *i.e.* the owner's right to sell it, is balanced in most legal systems with the state's right to buy the asset under specific circumstances where the public good outweighs the right of private property to protection from others.
- State ownership (public ownership): in many countries the most valuable biodiversity rich areas are owned and/or managed by the state. Restrictions on access and user rights, and the introduction of access or user fees, affect regular users of the resources. Corrective policies, for example giving rights to certain groups, compensation, voluntary agreements and leasing out some areas, can alleviate negative distributional effects.
- Community ownership or common property: the resource is managed by a group of users who can define the conditions of access to a range of benefits arising from the collectively used resource. This type of ownership structure is most likely to succeed when there is a relatively small group with shared needs and norms, clear boundaries for resource management, stability in the group undertaking management and relatively low costs of enforcement (Adger and Luttrell, 2000). Community ownership tends to be more equitable than other forms, while usually traditional groups define their own rights and manage the resource together. However the distribution of costs and benefits differs if the community ownership is introduced after an open access period or reintroduced after a state owned period. There is some empirical evidence that the management regime and the system of rights in some cases favour the richer households (Adhikari *et al.*, 2004).

There are particular challenges for understanding the institutional setting in which instruments are to be applied when property rights are not stable, but have recently changed; or, in some cases, are still evolving. Typical examples are transition economies (*e.g.* former socialist countries of Central and Eastern Europe). In such dynamic circumstances, policy instruments may have unforeseen consequences. A greater emphasis on prior information and ongoing monitoring is advised than for situations with more stable systems of property rights.

User rights are also important when discussing the distributive effects of biodiversity policies. There are a number of different rights to consider: rights of access, hunting or fishing rights, rights to collect nature-related goods, logging rights, development and housing rights and, in some cases, mining rights. The legal or informal systems governing these rights need to be taken into account because the introduction or modification of biodiversity policy measures can affect the previous holders of these rights and change their income options.

Property and user rights are historically determined and culturally embedded in different societies. These implicit and explicit systems of existing rights are a key consideration for instrument choice as different instruments alter these rights in very different ways and therefore give rise to immediate distributive effects.

Overview of biodiversity policy instruments and their distributional effects

While the establishment of protected areas, in particular national parks, is perhaps the most visible form of biodiversity policy, it is only one of a wide variety of instruments. Below is a non-exhaustive survey of the most important biodiversity policy instruments currently used:

- **Designation of protected areas:** land set aside (usually, but not always in state ownership) in a protected status with usually severe limitations on extractive and consumptive activities. Example: Yosemite National Park.
- **Land use regulations:** restriction on extractive and consumptive activities carried out on privately or commonly-owned land. Example: US Endangered Species Act.
- **Land purchases:** purchase of land for the purpose of conservation or habitat creation/maintenance from private or public landowners with no or highly restrictive use by the public. Example: Wild Rivers Legacy Forest, Wisconsin.
- **Conservation easements:** voluntary contracts between private landowners and conservation agencies specifying an exchange of payments or a tax credit for accepting land use restrictions. The price of the easement can be set by the government, negotiated between the contracting parties, or determined by auction. Example: BushTender programme in Australia (see Section 7.2.2).

- **Payments for ecosystem services:** voluntary contracts between providers and consumers of ecosystem services on a quasi-market basis. The providers will usually agree to manage their land in accordance with some restrictions on management practices. Example: PES in Costa Rica.
- **Control of trade in endangered species:** restrictions or ban of international movements of animals and animal parts. Example: the Convention on International Trade in Endangered Species of Wild Fauna and Flora (CITES).
- **Biodiversity-related taxes:** imposition of a general or hypothecated fee or tax on inputs or outputs considered harmful to biodiversity. Example: Danish pesticides tax.
- **Removal of perverse incentives:** reduction of fiscal or regulatory measures that reward activities harmful to biodiversity. Example: replacement in Austria of production-based agricultural subsidies with direct payments for environmental services (ÖPUL; see Box 3.2).
- **Market creation for biodiversity:** introduction of private property rights in biodiversity-related goods and services in combination with market-based trade. Examples: individual tradable quotas in fish catch in New Zealand; ecotourism.
- **Product certification:** provision of independent or state-authorised information certifying production in accordance with certain environmental criteria. Example: shade-grown coffee.

Obviously, the strength of this diversity of instruments lies in the fact that there is a plethora of tools for bringing about the desired policy result. Within this diversity, however, there are two key characteristics that influence the distributive impact of each instrument: *i)* participation and *ii)* the policy mode.

Participation and distributive effects. Instruments can be classified by the degree to which individuals interact with the policy:

- Voluntary policies allow potential participants to decide whether to contribute to the policy or not. Commonly, this means that the policy will only be able to recruit those potential participants who expect to do at least as well out of the policy as before. As a result, even if the potential losses are minute, a sufficient number of such potential losers can dramatically affect the uptake of a policy.
- Non-voluntary biodiversity policies, on the other hand, force individuals to participate in the policy. Typical policies are restrictions on property rights, *e.g.* by banning land development; or taxes and fees, *e.g.* a pesticides tax. The lack of choice about participation has two immediate consequences for the distribution of impacts. On the one hand, non-voluntary instruments

 PEOPLE AND BIODIVERSITY POLICIES – ISBN 978-92-64-03431-0 –

will – as a rule – produce losers at the individual level. As we will discuss in Part II, these welfare losses create incentives for losers to undermine the policy. On the other hand, being able to require mandatory participation dramatically enlarges the set of feasible policies for the policy-maker as policies do not have to ensure that every participant is at least not made worse off by participating. This gives policy-makers access to projects with the potential to generate significant net benefits at the aggregate level, but at the expense of small individual losers. Handling this trade-off between the creation of losers and viability of the policy in the face of losers' opposition is one of the key challenges in the design of conservation policies that Part III will explore.

Policy mode and distributive effects. The second influence of the distributive impact of an instrument is its policy mode. The policy can leave the existing ownership structure in place, but alter the rewards from carrying out certain activities. Reward-based policies leave it to the policy participant to decide how much of a certain activity is carried out, but specify a fee that typically increases with the volume of the activity being carried out. Fees can take the form of either a tax on an activity deemed damaging to the policy target, or a subsidy on an activity deemed to contribute to the policy target. Fees can also arise out of policies that use new or existing opportunities for excluding users, and this can turn some of the goods and services produced by biodiversity into private goods.

Welfare economics makes a clear statement about the difference in distributional impacts between instruments based on changes of property rights and those based on changes in rewards. Reward-based policies are in essence policies that change prices, and as such typically involve lower welfare losses than those based on, for example, restrictions on property rights. The reason is that participants faced with a change in rewards, but free from property rights restrictions, can shift production or consumption to, or away from, the targeted good or service to alternatives, depending on whether the reward is increasing or decreasing.

In the assessment of welfare impacts of the two modes of instruments, different measures become relevant. Equivalent or compensating variation are the appropriate measures for a change in reward (with change in consumer surplus a good proxy; see Willig, 1976). For changes in property rights, equivalent and compensating surplus are the correct measures (Just *et al.*, 2004).

The two dimensions of participation and policy mode result in four distinct types of policy instruments (Table 3.4). These are each discussed in turn below.

Table 3.4. **Classification of policy instruments**

Policy mode	Participation	
	Voluntary	Involuntary
Change in property and use rights	***Type II*** Land purchase Conservation easements	***Type IV*** Designation of protected areas Land use regulations Trade restrictions
Change in rewards	***Type I*** (Public) payments for ecosystem services Market creation Product certification	***Type III*** Biodiversity-related taxes User fees Removal of perverse incentives

Type I: Change in rewards and voluntary participation

Instruments in this category are characterised by two facts. One is that the individuals, households and groups participating in the policy are doing so voluntarily. The other is that the policy does not change ownership rights, but that the owners are now offered different rewards for certain activities. A typical example is payments for ecosystem services (PES). Landowners can decide freely as individuals or groups whether they would like to offer ecosystem services on the "market". Examples are market creation and product certification schemes, assuming that the costs of setting up these policies are insubstantial. Prior to the policy, offering such services would usually have been uneconomic as there were no contracts available involving reliable rewards. Once a PES is introduced, such services can be profitable. However, no landowner is obliged to offer such services.

It would be tempting to conclude that as this policy is voluntary, instruments of this type do not affect welfare negatively. The simple reason is that individuals are unlikely to voluntarily enter into transactions out of which they expect to incur a net loss. Relative welfare, however, will be affected because different parties accrue different gains from the policy. In particular, in the context of PES schemes and market creation, there is increasing evidence that parties at the lower end of the income and wealth distribution scale have fewer opportunities to realise the economic potential offered by these policy instruments than the better-off.

Furthermore, many instruments classified as voluntary, such as conservation contracts and PES, contain an element of coercion and hence there may be some losers. The reason is that these instruments commonly involve payments to the contracting parties, the funds for which have to be raised from some other groups. This involves a redistribution from the general public to voluntary participants via the general taxation system. Losers will therefore be those whose welfare gains from the policy's biodiversity increases

 PEOPLE AND BIODIVERSITY POLICIES – ISBN 978-92-64-03431-0 – © OECD 2008

are less than the welfare losses brought about by the increase in taxes they must pay to finance the policy. This feature is also found in Type II instruments (see Box 3.1).

Box 3.1. **Conservation easements in Colorado**

Throughout Colorado, local and state-wide land-protection organisations (land trusts) provide economic incentives for land conservation by offering landowners a state income-tax credit in exchange for a conservation easement on their property. An easement is an agreement, made between a landowner and a conservation organisation, which maintains the private ownership of the property while permanently prohibiting certain types of development.

Tax deductions for conservation easements are not new in the US. They have existed at federal level since 1976, and several states offer them. But Colorado is one of the most generous, remarkably so for a state not always known for caring for the environment. It offers an income-tax credit of 50% of the fair market value of the easement to a maximum of USD 375 000.

These credits are also impressively flexible. Income-tax credits may be fine for the Hollywood millionaires who own expansive tracts of land near mountain resort communities like Aspen and Vail; but they have limited appeal for the many Colorado ranchers and farmers who have little income for the state to tax. Such people, land rich but cash-poor, may now submit their credits to the state treasury for a full cash refund whenever the state budget is in surplus. Or (budget surpluses being rare in recent years) they may sell them, at between 80 and 85 cents to the USD, to a buyer who pays more in Colorado income taxes.

The legislation has had a massive impact: total area protected in Colorado increased from just under 142 000 hectares in 2000 to almost 1 million by the end of 2005. However, its cost in tax revenue is raising eyebrows. Annual revenue loss from the credits rose from a mere USD 2.3 m in fiscal year 2001, to USD 85.1 m in 2005.

Source: Adapted from *The Economist*, 2007.

In sum, if the policy objective passes the efficiency test, then the use of Type I instruments is least likely to concentrate losses in any particular individual or household. Instead, the distributive pattern will be one of concentrated benefits and small and highly diffuse losses, if any.

Type II: Change in rights and voluntary participation

Type II policies include conservation easements (see Box 3.1), the outright purchase of land for the purpose of conservation, and transfers of property rights from the state to, say, households or communities. Rather than providing incentives to landowners to behave in a certain way, this type of policy offers an exchange of property rights for some benefit (often cash), with the new owner managing the newly acquired rights as they deem fit within the confines of the policy. The rights that are being sold usually comprise all development rights for long time periods, sometimes indefinitely.

As for Type I policies, the voluntary nature of Type II policies implies that there would be no direct losers (other than those funding the policy through general taxation). Unless the price offered for the right to be surrendered exceeds its valuation by the current owner, there will be no participation and therefore no loss. Where that price exceeds the valuation, the previous owner will be better off accepting the exchange.

Empirically, there is no evidence of losers from Type II policies. However, there is evidence that some gain more than others. This evidence comes from voluntary systems involving the sale of property rights for conservation purposes pioneered in the agricultural context in industrialised countries. Monitoring of the two large land preservation programmes in the US, namely the purchase of development rights or agricultural conservation easements (PDR/PACE) and the transfer of development rights (TDR) programmes, shows that the better-off tend to benefit more. Those benefiting from these programmes are likely to have more assets, likely to be farming land extensively, and likely to rely more on farming for their income (Lynch and Lovell, 2003). This pattern is by no means unusual: there are similar effects in other countries where perverse subsidies have been replaced by voluntary payment schemes (see Box 3.2).

There are similar effects from Type II instruments which transfer property rights to local communities. If these communities or households are able to manage the biodiverse habitats and ecosystems better than the government, there is no loss involved for the general public. However, as before, the extent to which different groups benefit can vary. Adhikari *et al.* (2004) studied the distributional effects of property rights devolution from central governments to community-based local users groups in Nepal initiated in the 1990s. They found that while such policies are successful in halting deforestation and enhancing biodiversity conservation, the distributional effects within local user groups tend to be highly uneven and accentuate pre-existing distributional patterns. Poorer households in many cases had at least the same access to the communally managed resources after devolution as before, but less access than those better off. This, along

 PEOPLE AND BIODIVERSITY POLICIES – ISBN 978-92-64-03431-0 – © OECD 2008

Box 3.2. **Differential impacts of ÖPUL on crop farmers and livestock farmers**

ÖPUL is an agro-environmental scheme established in Austria in the early 1990s. Its aim is to replace agricultural subsidies based on the volume of production with direct payments for environmental services (OECD, 1999). Many of these environmental services are measures intended to safeguard and improve the biological diversity of the cultivated landscape, with which many species have co-evolved. Examples of these services include maximum limits on livestock numbers, crop rotation, set-aside and specific mowing patterns. Farmers are offered a menu of farming practices from which they can voluntarily choose the appropriate measures. Once signed up to the scheme, farmers receive area based payments for each measure. Evaluations of the socioeconomic effects of these policies between 1998 and 2002 have demonstrated two important distributional impacts. The first is functional: replacing rewards based on intensive production with incentives for extensive practices led to a policy inherently biased towards crop farmers. Land area-based payments thus led to a redistribution away from livestock farms and processors. The second is a scale impact: larger farms have been able to benefit considerably more from the new policy than smaller farms in terms of payments received (Groier, 2004).

with other studies (*e.g.* Campbell *et al.*, 2001), challenges the notion that property rights devolution is an unqualifiedly successful instrument for reconciling conservation and pro-poor policy objectives.

In all, Type II policies involve rather benign distributive effects and lead in general to a redistribution from the general public to participants, in most cases through general taxation or through relinquishing previously publicly held property rights. At the origin, the distributive impacts of such instruments are therefore the same as for any other public project, and depend on the characteristics of the fiscal system. This means that, due to the progressive nature of the taxation system in most countries, the distribution of costs for providing voluntary policies have a progressive impact on the population.

Type III: Change of rewards and involuntary participation

Type III instruments combine coercive participation with a change in rewards. The coercive element usually arises out of particular powers vested in the state, such as fiscal powers. Taxes and user fees are therefore typical of this type.

Coercion enables the state to target those groups benefiting from biodiversity-related resources and to impose on them the financial burden of a policy. Careful analysis and discussion with the affected groups can help set the right level of incentive. The distributive effects depend critically on the design of the tax or fee scheme, both at source and at destination. At source, policy-makers can target distributive outcomes through selective exemption or inclusion of certain activities in the scheme. For example, research into park admission fees indicates a clearly progressive effect (Feinerman *et al.*, 2004), while taxes on agricultural machinery are likely to affect low income households. At the destination, policy-makers can choose to earmark tax revenue for a clear group of beneficiaries or choose to allocate tax revenues to the general budget. For example, revenues from the Danish pesticides tax, which predominantly affects conventional arable crop producers, are partly used to support organic agriculture (Schou and Streibig, 1999).

In sum, therefore, Type III instruments offer considerable scope for influencing the distributional effects of biodiversity policies while retaining attractive economic features. This explains the popularity of this instrument type in redistributive policies in other contexts (Serret and Johnstone, 2006).

Type IV: Change of property and use rights and involuntary participation

Type IV encompasses many of the most commonly practised forms of biodiversity policies. These take the form of either outright takings, *i.e.* when the government forces landowners to give up formal and informal property rights in land that is to be turned into a protected area, or of restrictions on private property rights. When coercive restrictions on private property rights are sufficiently substantial, they are at times referred to as "regulatory takings". Depending on the specific circumstances of the case, classification as a regulatory taking may give rise to compensation for the original owner of the property right.

The extent to which access restrictions lead to adverse distributive effects clearly depends on a number of exogenous factors, as well as the policy design. Important exogenous factors are the availability of alternative livelihood opportunities (Wells *et al.*, 1992) and the interaction of the policy objectives with other determinants of local welfare. Design features include the channelling of conservation benefits to the local population through local employment in parks, through the encouragement of ecotourism and through a reduction of costs to the local population (Heady, 2000). These accompanying policy components can in theory have important mitigating impacts. At the same time, there is considerable debate about the actual capacity of these policy components to soften or even reverse primary impacts. While some researchers point to cases where a well-designed ecotourism initiative coupled with revenue-maximising entrance fees should

 PEOPLE AND BIODIVERSITY POLICIES – ISBN 978-92-64-03431-0 –

generate net positive impacts at the local level (Naidoo and Adamowicz, 2005), meta-analytical studies suggest that ecotourism cannot reconcile sustainable conservation with sufficient tourism income (Krüger, 2004).

The most radical form of intervention in biodiversity policies is that of outright takings. Takings, also referred to as "eminent domain" in the US context, or compulsory purchase in the UK, allow government to force owners of land to give up – without their consent – formal and informal property rights in land that is to be turned into a protected area. This practice goes back to the very beginnings of nature conservation policies, such as the establishment of the first national parks in the US (Burnham, 2000). A common feature of protected areas is to ban extractive, productive, consumptive and non-consumptive use by individuals who previously enjoyed the benefits of this use. A recent survey of 138 projects receiving financial assistance from the World Bank, and in particular from the Global Environmental Facility, found that 120 restricted access by previous users.

The extent to which these types of property rights restrictions result in actual changes in use by previous users depends on the level of enforcement, for example by guards (Bruner *et al.*, 2001) or other forms of policing (Albers and Grinspoon, 1997). When enforcement is carried out, protected areas can be effective, particularly in reducing extractive pressure (Bruner *et al.*, 2001). Reduction of productive pressure has been less effective, both at the site-specific and at the aggregate level (Bruner *et al.*, 2001; McNeely and Scherr, 2003).

The compensation of previous owners in such cases is key and requires the value of land to the previous owner to be determined. The distributive impacts then depend on the process used to determine the proper form (monetary or non-monetary) of compensation and the distributive impact of raising the public funds necessary to provide the compensation.

Distinct issues arise in cases where takings concern settlements and habitations and where the designation of land use stipulates the absence of occupancy or use. As a result, takings give rise to the displacement of former owners and – in the case of non-compliance – to their eviction. Again, displacement is associated with the very earliest conservation policies. The creation of the US national parks frequently involved displacement of indigenous populations (Spence, 1999). Sizeable resettlement programmes are a component of most African national parks, such as the Korup National Park in Cameroon (Schmidt-Soltau, 2003), the South African national parks established under apartheid (Carruthers, 1995), and protected areas in Tanzania (Chatty and Colchester, 2002). In Asia, Indian conservation policy has involved several well-documented resettlement programmes

(Rangarajan, 1996; Saberwal *et al.*, 2000), as has the Royal Chitwan National Park in Nepal (McLean and Stræde, 2003).

The welfare impacts of resettlement on individuals are complex and multi-faceted, ranging from income losses as a result of access restrictions and loss of productive assets, to less tangible but no less important psychological costs. Empirical estimates of the economic effects of resettlement are patchy (Table 3.5).

Table 3.5. **Income loss estimates as effects of resettlement**

Name	Total area in km^2	Population (approx.)	Estimated annual income loss from hunting/gathering in euros		
			Per capita in cash	In cash	Total
Dja Biodiversity Reserve	5 260	7 800	69.82[a]	544 596	956 103
Korup National Park	1 259	1 465	76.0[a]	111 369	195 522
Lake Lobeke National Park	2 180	4 000	69.82[a]	279 280	490 309
Boumba Beck National Park	2 180	4 000	69.82[a]	279 280	490 309
Dzanga-Ndoki National Park	1 220	350	69.82[a]	24 437	42 902
Nsoc National Park	5 150	10 000	69.82[a]	698 200	1 225 772
Loango National Park	1 550	2 800	69.82[a]	195 496	343 216
Moukalaba-Doudou National Park	4 500	8 000	69.82[a]	558 560	980 618
Ipassa-Mingouli	100	100	69.82[a]	6 982	12 258
Cross-River NP Okwangwo	920	2 876	158.96[a]	457 169	802 614
Nouabalé Ndoki National Park	3 865	3 000	69.82[a]	209 460	367 732
Odzala National Park	13 000	9 800	69.82[a]	684 236	1 201 257
Total/average	**41 384**	**54 000**		**4 049 065**	**7 108 612**

a) Estimated value, see Table 3 in source.
Source: Cernea and Schmidt-Soltau, 2006.

There are two important features to remember when considering resettlement policies in protected areas. The first is that in many protected areas, presence of resident populations is the norm rather than the exception. Bruner *et al.* (2001) reported that, globally, 70% of tropical protected areas with a ban on consumptive uses contain resident populations and 54% contain residents who dispute the ownership arrangements in at least some of the park's land. These numbers are similar to South American parks,[3] the Eastern Kalimantan region of Indonesia (Jepson *et al.*, 2002), the Gobi desert of Mongolia (Bedunah and Schmidt, 2004) and protected areas in Myanmar (Rao *et al.*, 2002).

The second feature is that many resettlement programmes fail in practice due to local resistance. The relationship between the original policies and the

PEOPLE AND BIODIVERSITY POLICIES – ISBN 978-92-64-03431-0 – © OECD 2008

post-implementation situation is a function of the level of enforcement. In situations where enforcement is imperfect, the *de facto* changes imposed on local populations are often less severe than those foreseen under the policy on paper. Observations that a full-scale implementation of the Indian protected area policies would result in up to 4 million displaced people (Kothari, 2004) and that full enforcement of protected areas in Africa could involve a similar magnitude of displacement (Geisler and de Sousa, 2001), give an idea of the scale of possible eviction and the likely obstacles to the full implementation of these policies.

In conclusion, the relationship between instrument choice and distributive impacts shows that policy-makers have considerable scope for assigning the benefits and costs to different groups depending on what instruments they choose in a specific context. The potential for mitigating the primary distributive effects of a policy in a second round through clever instrument choice is therefore significant. However, there are trade-offs in instrument choice between the desirability of being able to fully implement policy (which calls for coercive instruments) and the desirability of being able to avoid creating a high volume or a high individual incidence of policy losers (which calls for voluntary instruments). Historically, this trade-off has led to a strong bias towards policies that combine coercion with changes in property rights. The result has often been unfair distributive outcomes and a failure of the policy on the ground. Other approaches, such as tax-based measures, seem underexploited in their potential to strike a middle ground in this trade-off.

3.2.3. Spatial dimension

Spatial patterns are of critical importance in understanding the challenges inherent in biodiversity policies. The first aspect is the relationship between areas of human use and areas of high conservation priority. Balmford *et al.* (2001) mapped the spatial coincidence of areas of high conservation importance and areas of high primary productivity in Africa. They found a strong positive correlation, indicating that land use conflicts between land conversion for development and land preservation for conservation are the norm and will increasingly surface. Luck *et al.* (2004) found a high spatial correlation between areas of high species richness and areas of high human population density in Australia and North America, reaching similar policy conclusions. This general pattern has been confirmed in more specific settings by Gaston (2005), highlighting that the spatial distribution of humans and biodiverse habitats and ecosystems gives rise to conflicts.

The second general spatial pattern of importance is the relationship between areas of low economic development and areas of high conservation priority. Angelsen and Wunder (2003) established a strong link between centres of high biological diversity and endemism and low income and low asset endowments. Cavendish (2000) reached similar conclusions, as did Markandya (2001). At a global level, the correlation is even more palpable, with industrialised countries characterised as "income rich and biodiversity poor" and developing countries as "income poor and biodiversity rich" (Swanson, 1996).

As a result, the implementation of biodiversity policies has a critical spatial dimension, raising questions both of international equity between countries and between groups of very different economic levels. These questions are salient because the primary costs of implementing biodiversity policies are frequently locally concentrated in those areas where biodiversity is to be maintained and enhanced. These costs are borne by populations whose access to benefits generated in the area where these policies are implemented is restricted in part or completely as a result. At the same time, many of the benefits generated by policies that maintain and enhance biodiversity are spatially diffuse, arising often at locations many hundreds or thousands of miles away and to individuals or groups that ultimately depend less on the area to be protected than those people living near or within them.

Benefits and costs in a spatial context: local, national and international dimensions

The spatial dimension is most palpable in non-voluntary, quantity-based instruments such as the establishment of protected areas. In a seminal paper, Wells (1992) studied this geographical pattern of distribution of costs and benefits. Building on Dixon and Sherman's (1991) influential study on the benefits and costs of protected areas, Wells demonstrated the presence of a "spatial mismatch". The nature of this mismatch can be shown with the help of a simple spatial accounting matrix that examines the intersection between different spatial scales (local, regional/national, and transnational/global) on the one side, and different benefit categories (Table 3.6) and cost categories (Table 3.7) on the other.

Looking at Table 3.6, Wells' key observation is the limited scale of benefits at a local level; benefits rise to intermediate levels at the regional/national scale and become of major significance at the transnational/global level. While many of these observations were based primarily on intuition and common sense, they have since been substantiated by empirical evidence. A typical example are studies on consumers' willingness to pay for the conservation of biodiversity in some other spot on the planet, frequently with little likelihood that they would ever visit this location or derive some personal direct benefit (see Kramer and Mercer, 1997).

 PEOPLE AND BIODIVERSITY POLICIES – ISBN 978-92-64-03431-0 – © OECD 2008

Table 3.6. **Relative significance of protected area benefits on three spatial scales**

Protected area benefits	Spatial scales		
	Local	Regional/national	Transnational/global
Consumptive benefits	*0-3*	0-2	0-1
Recreation/tourism	*0-3*	*0-3*	0-1
Watershed value	0-2	*0-3*	0-1
Biological diversity	0-2	1-2	*0-3*
Nonconsumptive benefits	0-2	0-1	*1-3*
Ecological processes	1-2	1-2	2-3
Education and research	0-2	0-1	2-3
Future values (for all of the above categories)	*0-3*	*0-3*	*0-3*

1. 0 = insignificant, 1 = minor significance, 2 = moderate significance, 3 = major significance.
2. The underlined figures are at the scale where the benefit category has the potential to be most significant.

Source: Wells, 1992.

This is not to say that there are no local or regional benefits from biodiversity conservation. Frequently, biodiversity policies help resolve co-ordination and co-operation failures among current local users and can thus create pro-poor benefits from protected areas (Alix-Garcia *et al.*, 2004). A very significant proportion of benefits, however, accrue far from the conservation location.

Wells (1992) found that the spatial incidence of the various costs of conservation tends to show a geographical pattern that is the inverse of that for benefits (Table 3.7).

Table 3.7. **Relative significance of protected area costs on three spatial scales**

	Spatial scales		
	Local	Regional/national	Transnational/global
Protected area costs			
Direct costs	0-1	*0-3*	0-1
Indirect costs	*0-3*	0-1	0-1
Opportunity costs	*0-3*	*0-3*	0-1

1. 0 = insignificant, 1 = minor significance, 2 = moderate significance, 3 = major significance.
2. The underlined figures are at the scale where the cost category has the potential to be most significant.
3. Protected area cost categories from (6).

Source: Wells, 1992.

Netting out benefits and costs for each geographical level (Table 3.8) gives rise to the diagnosis that there is a spatial mismatch between the level at which the most significant costs are borne (local and national) and the level at which the most significant benefits arise (global).

Table 3.8. **Spatial mismatch of potentially most significant costs and benefits**

Potentially most significant benefits (from Table 3.6)	Potentially most significant costs (from Table 3.7)
LOCAL SCALE	
Consumptive benefits	Indirect costs
Recreation/tourism	Opportunity costs
Future values	
REGIONAL/NATIONAL SCALE	
Recreation/tourism	Direct costs
Watershed values	Opportunity costs
Future values	
TRANSNATIONAL/GLOBAL SCALE	
Biological diversity	(Costs minimal)
Nonconsumptive benefits	
Ecological processes	
Education and research	
Future values	

However, this diagnosis of a spatial mismatch is clearly predicated on the case of protected areas. As we have seen above, these instruments, though the most effective on paper (but perhaps not in practice, see Cernea and Schmidt-Soltau, 2006), have the most pronounced welfare impacts. It is therefore not surprising that these results would arise.

The spatial mismatch hypothesis is both supported and challenged by recent detailed research. De Lopez (2003) examined the distribution of costs and direct use benefits of three different management scenarios in the Ream National Park in Cambodia. The costs and benefits were examined for four different groups of stakeholders: the population resident within the park (local impacts); commercial entrepreneurs (regional/national impacts); the armed forces (national impacts); and visitors (international impacts). Local residents rely on the park area for firewood, medicinal plants, timber and non-wood forest products (at an annual value of roughly USD 160 000) and fish. Different park regimes impose no or severe restrictions on these activities for locals and outsiders and combine these with alternative income from tourism-based revenues. The analysis showed that in terms of total net benefits created, conservation-focused policies were only marginally better than more extraction-focused policies. However, in contrast to Wells (1992), the distributional consequences of conservation-focused policies were found to be pro-poor. The reasons are: *i)* that local residents gain from an increase in visitor numbers; and *ii)* more of the existing use benefits are reserved for locals since restrictions make it less attractive for outside entrepreneurs and the army to compete for fish and timber resources.

This conclusion mirrors one of the ambiguities in the establishment of protected areas at the national level and challenges the simple logic of "spatial mismatch". Protection for conservation can be pro-poor because protected areas spell out – often for the first time – the precise nature of use rights in an area. Local communities may gain from protected areas because outside competitors are excluded. Typical examples are the creation of the Kakadu National Park, which local residents considered as a protection against the threat of uranium mining in the area (Lawrence, 2000); the Gates of the Arctic National Park in Alaska, which local residents considered as a protection against the installation of an oil pipeline (Catton, 1997); and also most of the extractive reserves within the Brazilian Amazon, which were thought by indigenous people to offer protection from outside settlers and loggers (Goeschl and Igliori, 2006).

International mechanisms and distributive effects

At the international level, there is a concern to manage the global aspects of biodiversity provision and the distribution of benefits and costs from these activities. This gives rise to a number of issues of distributive importance, each discussed below:

- The potential WTP in developed countries for biodiversity conservation elsewhere and its relation to actual funds.
- The impact of global funds for biodiversity on recipient countries.
- The global rules devised for sharing the benefits from international co-operation on biodiversity protection and their distributive impact among countries.

Willingness to pay for biodiversity conservation. One international issue is the demand in some countries for the conservation of biodiversity in others. Recent years have seen the first attempts to provide quantitative estimates of the willingness to pay in developed countries for biodiversity to be protected elsewhere. Kramer and Mercer (1997) conducted a mail survey of a random sample of 1 200 US residents, including questions to gauge knowledge of and attitudes towards rainforest conservation, socioeconomic status and willingness to pay (WTP). They then used contingent valuation methods to gauge the WTP to double the amount of terrestrial national parks and nature reserves in tropical countries. This recorded a total willingness to pay (for all US households) of USD 2.18 billion, based on a mean WTP per household of USD 24.

Horton *et al.* (2003) surveyed 407 residents in the UK and Italy, and used contingent valuation to gauge their WTP for an expanded national parks system that would cover 5% of the Brazilian Amazon. The total willingness to pay was USD 1.8 billion, based on a mean WTP per household of USD 45. Perhaps even more significantly, the authors asked respondents about their equity position on shouldering the cost of tropical rainforest protection. They reported that 93% of respondents believed that the industrial nations should share some of the costs of tropical rainforest protection (51% of the total cost).

There are several international mechanisms used to capture industrialised countries' willingness to pay for protecting biodiversity in developing countries. These include the conservation programmes of various international donors such as the European Union, World Bank and the Global Environment Facility (discussed in Chapter 7).

Domestic impact of biodiversity funding. A second issue is the domestic impact in countries that receive international funds for biodiversity protection. Countries agreed in the early 1990s to create international institutions for the provision of this global public good and its services. The Convention on Biological Diversity (CBD) in 1992 and its financial instrument, the Global Environment Facility (GEF), are the primary international agreements for implementing co-operation for this global good.[4]

International distributive rules. Whilst the creation of global institutions such as the CBD reflects the international policy-making community's desire for global co-operation in the conservation of biological diversity, there are important distributional issues in this co-operation. These arise from the substantial asymmetries that exist across the countries concerned. Some parts of the world are highly endowed with biodiversity (here, the "South"), while others have very little (the "North"). The North is relatively well-endowed in human and physical capital, giving rise to relatively high average incomes and levels of wealth. The South, on the other hand, contains few of these sorts of resources and assets. These asymmetries result in an unbalanced bargaining process, with each party negotiating from a position of relative strength in some respects, and weakness in others. But co-operation in combining the endowments of the North and the South could yield significant welfare gains for both societies, so long as both can agree how these gains will be distributed.

The contractual terms of the Convention of Biological Diversity can help divide these gains (Gatti *et al.*, 2004). These terms include a framework within which the North and South agreed that:

- A country's biodiversity is under its sovereign control.
- Biodiversity should be provided as a global good.

- A contractual basis would be created for determining how states providing biodiversity should share in the benefits of the global public good they provide.

In other words, the South has full property rights in its biological diversity and can devise policies for its domestic management without outside interference. But the South must also provide much of the biodiversity resource in exchange for compensation: "[the North] shall provide new and additional financial resources to enable [the South] to meet the agreed full incremental costs to them of implementing measures which fulfil the obligations of this Convention" (Art. 20, CBD, 1992).

The meaning of the term "incremental costs" is further defined within the GEF as:

> [the costs of] additional national action beyond what is required for national development [the baseline] that imposes additional [or incremental] costs on countries beyond the costs that are strictly necessary for achieving their own development goals, but nevertheless generates additional benefits that the world as a whole can share... (GEF/C.7/Inf.5: para.2 and GEF/C.2/6 para.2)

These terms of the agreements establishing the CBD and the GEF specify a very particular distribution of the net surplus from international co-operation on biodiversity. They contain an obligation on those states hosting biodiversity to shoulder the cost of supplying it for the global good, and dictate that the North shall share the benefits of such public goods with the South by paying the amounts required to compensate it for the costs of its participation.

However, the CBD has so far failed to achieve distributive goals. With most of the arguments above pointing to domestic government failures, the question becomes why the international agreements do not induce the governments rich in biodiversity to correct these distortions. One explanation is that this has to do with the distributive arrangements within the terms of the convention. Gatti *et al.* (2004) showed that under the terms of incremental cost, countries may have little incentive to invest in the conservation of biodiversity when compensation for the global benefits generated is only paid for marginal improvements.

The CBD example shows that the institutions governing the creation of its surplus (concerned with efficiency) and its division (concerned with distribution) can be interlinked, particularly in a setting where no higher authority exists to initiate, monitor and enforce conservation outcomes. We return to these aspects in Part II.

Notes

1. See Russell and Vaughan (1982) for a study on the relationship between income and the demand for recreational fishing.
2. A third case of increasing benefits from each addition to biodiversity is not considered here.
3. 85% of which have people living within their boundaries (Amend and Amend, 1995).
4. In the specific area of genetic resources there is an additional emerging international instrument – the International Treaty on Plant Genetic Resources for Food and Agriculture. But its future impact is unclear since: *i)* important countries (*e.g* the United States, China and Brazil) have not signed or ratified it; and *ii)* key mechanisms of the treaty, (*e.g.* the financing mechanism) have not been defined.

PEOPLE AND BIODIVERSITY POLICIES – ISBN 978-92-64-03431-0 – © OECD 2008

ISBN 978-92-64-03431-0
People and Biodiversity Policies
Impacts, Issues and Strategies for Policy Action

Chapter 4

The Distributive Effects of Biodiversity Policies: Dynamic Analysis

4.1. Intergenerational equity: evaluating costs and benefits across time

Biodiversity policies have an explicit time dimension. The total economic value of biodiversity concept contains some important intertemporal components, such as option values, exploratory values or quasi-option values (Bulte and Withagen, 2006), and bequest values (Pearce and Moran, 1994). These values are conceptually tied to the future in the following ways:

- Option values arise from the continued preservation of habitats and ecosystems by allowing conversion decisions to be postponed into the future.
- Quasi-option values arise from new information becoming available in the future that allows new decisions about biodiversity management to be made.
- Bequest values arise from the ability to *pass on* habitats and ecosystems to future generations.

The time scales over which these value components become relevant range from the very short to the very long. Progress in the life sciences may very rapidly allow the identification of valuable genetic traits in plants or micro-organisms that cannot currently be identified. The time scale of the exploratory or quasi-option value may therefore be measured in months or a few years. On the other hand, in the context of climate change, society may want to postpone decisions to irrevocably convert habitats given that it is not clear at the moment to what extent habitats may be degraded as a result of changing precipitation patterns. Here the time scale would appropriately be measured in decades or rather hundreds of years.

Given this explicit time dimension, policy decisions today will affect individuals currently alive, as well as generations not yet born. The policy-making process therefore needs to compare benefits and costs of biodiversity policies that may arise at vastly different points in time and justify them against some measure of intergenerational equity. Methods to do this are commonly referred to as discounting techniques.

Discounting is a major concern for intergenerational equity in biodiversity policies: the longer the time horizon of the effects of a specific policy, the larger the impact of discounting. Hence, the evaluation of policies involving irreversible components, such as species extinction, loss of habitats and ecosystems depends

 PEOPLE AND BIODIVERSITY POLICIES – ISBN 978-92-64-03431-0 – © OECD 2008

to a large extent on the choice of the discounting model and its parameters. While this book can only broadly cover the issue of discounting, more detailed but accessible treatments (though not specifically on biodiversity policies) can be found in Pearce *et al.* (2003) and Groom *et al.* (2005).

4.2. Discounting

The economic tool for project evaluation is cost-benefit analysis (CBA). The basic rule is that if the social benefits exceed the social costs of a particular policy then it increases social welfare and should be implemented. This is a straightforward concept if both costs and benefits occur at the same instant or at least within a reasonably short time period, *e.g.* logging a single tree to obtain firewood in an otherwise intact forest. The costs of logging and the benefits of consumption occur in close succession.

However, if there is a considerable time interval between the two, for example if the costs have to be incurred right away while the benefits occur at some stage in the future (an investment), then how do we compare flows at such different points in time? Box 4.1 explains discounting. Extending the above example, if the tree is felled to build a house rather than to heat it, then the benefits occur over a longer period, *i.e.* the lifespan of the house. On the other hand, if the forest from which the tree is taken is close to collapse, the logging might have a significant impact on the survival of this ecosystem and the species therein. Hence, the costs are long-term too and might even contain irreversible elements, *e.g.* if some species are unique to this forest.

Box 4.1. **Discount factors**

Discounting is a method that systematically assigns different weights, called *discount factors,* to costs or benefits occurring at different points in time. These weights decrease over time, rendering distant costs and benefits less important. The conventional form of discounting, called "exponential discounting" uses a constant *discount rate* (s) to calculate discount factors w_t. The appropriate formula is then:

$$w_t = \frac{1}{(1+s)^t}$$

In principle, it is clear how individuals deal with the problem of assigning weights to future flows (*e.g.* payments) and there is a sizeable theoretical and empirical literature on how individuals discount future payoffs (Frederick *et al.*, 2002). However, it is very much debated how society as a whole should value costs and benefits occurring at different points in time and to different generations. Individuals usually prefer benefits now to benefits in the future, and benefits enjoyed by their children to benefits enjoyed by their great-grandchildren. This applies to money as well as to risks (Pearce *et al.*, 2003). Hence, if people's preferences count in policy evaluation, this impatience should show up in the cost-benefit rule.

In contrast to this, philosophers and prominent economists have argued in favour of a zero social discount rate (Broome, 1992; Ramsey, 1928; Solow, 1974) whereby the current generation should receive the same weight as all generations to come. One reason for discounting usually acknowledged by this school of thought is the fact that in each period there is a positive but very small probability that the human race will become extinct (Stern, 2006), perhaps by the impact of a meteorite or a highly infectious and deadly disease for which no antidote is found. Hence, there is a chance that future generations might not exist and hence any cost or benefit occurring to them can be discounted accordingly.

But even if present and future generations are treated equally, a further reason for discounting consumption (rather than utility) is that future generations are likely to be better off than current ones. Hence, given a decreasing marginal utility of consumption, an additional unit is worth less according to the future generation's own preferences than to the current ones. This effect competes with a contrasting one running in the opposite direction: there are some goods, such as many environmental amenities and biodiversity used for bioprospecting, whose availability does not increase at the same speed as consumption of manmade goods and for which no close substitutes are available. The marginal utility derived from such goods increases over time. This effect is reinforced if the supply of such goods declines due to conversion of natural landscapes, biodiversity loss and environmental degradation (Krutilla, 1967).

The probability that the human species will become extinct in any period is (by orders of magnitude) smaller than discount rates deduced from individuals' behaviour (see Frederick *et al.*, 2002). Moreover, in the latter there is a gap between developed and developing countries. While for the former, discount rates below 10% are common, for the latter, values above 25% and even above 100% have been estimated (GEF, 2006), reflecting the specific planning conditions in developing countries such as lower life expectancy, liquidity constraints and lower security of property rights. Hence, if policy choice is to be based on individual preferences, it might be crucial whether future costs and benefits occur to individuals living in developed or in developing countries. Moreover, applying discount rates based on empirical evidence in developed countries (*e.g.* 3.5% for the UK, see HM Treasury, 2003) can result in a serious lack of acceptance by local stakeholders in developing countries. If the benefits of the policy occur in the future, stakeholders might

 PEOPLE AND BIODIVERSITY POLICIES – ISBN 978-92-64-03431-0 – © OECD 2008

put a considerably lower value on them than the planner assumes. This is an important constraint for the design of voluntary and incentive-based biodiversity policies.

How the choice of the discount rate matters in the medium and long run is illustrated by Figure 4.1. It presents the evolution of discount factors corresponding to different discount rates. Discount rates of 0, 0.01 (1%), 0.02 (2%), 0.05 (5%) and 0.1 (10%) per year are shown over a 200 year period. The lines show what happens to an initial quantity (w = 1 in year 0) as a result of those discount factors. So, for people with a discount factor of 2%, a promise of EUR 100 in 40 years is today worth only EUR 45 (w = 0.45 in year 100). Alternatively, if a foreseen event was to cause a loss of EUR 100 in 40 years, then we would only be willing to pay EUR 45 to avoid that loss. In other words, a predictable loss of EUR 100 40 years from now might go unmitigated.

Figure 4.1. **The evolution of the discount factor over time for different constant discount rates**

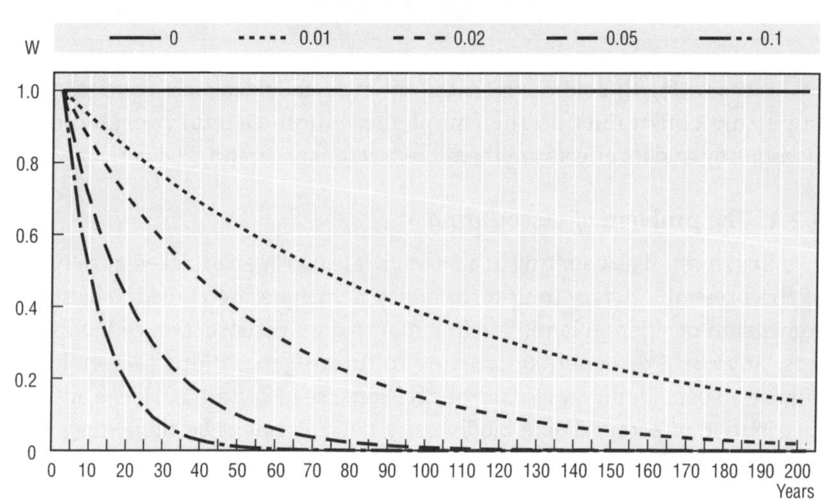

What is the effect of discounting on decisions? By attaching lower values to costs and benefits occurring in the distant future, discounting (and hence, the choice of s) has a major impact on the outcome of cost-benefit analysis and project evaluation. This is especially so for distributive effects between generations where, by definition, long time horizons are involved and thus discounting is a key determinant for identifying a desirable policy. Higher discount rates imply lower importance attached to future generations. To illustrate this point, consider the examples in Table 4.1.

A biodiversity conservation project costs 1 million right now but yields 5 million in conservation benefits 50 years in the future. Whether the project passes the cost-benefit test depends on the discount rate chosen (see Table 4.1). For s = 2%, the discount factor for t = 50 is 0.3715. Hence, the net benefit is *NB* = 0.3715 x 5 m – 1 m = 0.8575 m and is thus greater than zero. The project is beneficial at a social discount rate of 2%. Repeating the same exercise with s = 5% yields a very different result. The corresponding discount factor is 0.0872 ($w_{50}[s=5\%]$). The net benefit is negative (*NB* = 0.0872 x 5 m – 1 m = –0.564 m). Hence, at a 5% social discount rate, the project would not go ahead.

Table 4.1. **Two hypothetical cost-benefit scenarios with exponential discounting**

Costs		Benefits		Discount rate *s*	Discount factor *w*	Net benefit	Evaluation
Amount	Year	Amount	Year				
1 million	0	5 million	50	*2%*	0.3715	0.8575 million	Desirable
1 million	0	5 million	50	*5%*	0.0872	–0.564 million	Not desirable

Discounting therefore affects the set of socially desirable policies, as well as putting constraints on their implementation. Careful consideration of discounting and its effects are therefore key for successful biodiversity policies.

4.2.1. *The problem of discounting*

Exponential discounting at a positive rate has been attacked as a "tyranny of the present". If very long-term policy decisions are considered, such as conversion of pristine land, flooding due to dam construction or biodiversity loss, any costs or benefits occurring to future generations receive little to almost no consideration in current decisions (see Figure 4.1). Hence, although distributive effects of some biodiversity policies might be huge, they would frequently be dwarfed by the application of discounting.

The immediate relevance of the discount rate to biodiversity policies is a mainstay of natural resource economics: Clark (1973) demonstrated that high intertemporal discount rates are a key reason why many managed species have been "rationally" driven to, or close to, extinction in the past, because the future benefits of their existence have been considered negligible when decisions were taken. Swanson (1994) extended Clark's (1973) framework to species and habitats that are not managed, and showed that the same logic can also explain habitat conversion and deleterious management practices that give rise to "extinction by neglect". Discounting and sustainable development are often regarded as irreconcilable.

The most popular proposal for solving the "tyranny of the present" is to abandon discounting altogether (*i.e.* using a zero discount rate). This

 PEOPLE AND BIODIVERSITY POLICIES – ISBN 978-92-64-03431-0 – © OECD 2008

essentially gives the same weight to all current and future generations, including those living a million years from now. Hence, any project that at some stage yields benefits that are greater than the costs (both undiscounted) is worthwhile. However, zero discounting, taken seriously, has important implications for broader macroeconomic decisions such as the savings rate. Savings should by far exceed their current level and consumption by the current generation should fall considerably in order to yield high benefits to a generation in a far distant future. In fact, current consumption might conceivably be at risk of being driven down to subsistence levels. On a more abstract level, zero discount rates also raise the possibility that it is no longer possible to even formulate an optimal consumption and savings path (Koopmans, 1965; Asheim *et al.*, 2001). Put somewhat more pointedly, using a zero discount rate might have prevented mankind from converting any pristine land and from using any non-renewable resources. Zero discounting has therefore been labelled "tyranny of the future".

4.2.2. Intergenerational equity: the role of uncertainty and risk

In terms of intergenerational equity, both a constant positive discount rate and a constant zero discount rate lead to unsatisfactory policy outcomes over the long time scales that characterise many biodiversity policies. So does a balanced solution exist?

One hopeful candidate is "hyperbolic discounting" (Box 4.2). While using strictly positive discount rates, it differs from exponential discounting in one important respect. The discount rate s is not constant but decreases over time. Hence, discount factors decrease less than they would for constant discount rates in the long-run.

One major argument in favour of declining discount rates is uncertainty about future states of the world. Two conceptualisations of this uncertainty have been proposed. While Weitzman (1998) assumes uncertainty over the future discount factor, Gollier (2002a, b) allows for uncertainty over consumption paths. Both approaches come to the same conclusion: discount rates are declining. Uncertainty over future states of the world is common in biodiversity policies. The bioprospecting value of species, the effect of losing ecosystem services and the preferences of future generations are all highly uncertain.

Declining discount rates are backed by a very different recent theoretical approach in the social choice tradition. If a social planner advocates a mixed goal which combines a high discount rate and a low (zero) discount rate, the result is a social discount rate that declines over time (Chichilinsky, 1996; Li and Löfgren, 2000). Moreover, hyperbolic discounting is supported by empirical evidence (see Frederick *et al.*, 2002); people seem to apply hyperbolic discounting in their everyday decision-making.

However, there is a drawback to the concept of declining discount rates. Most types suffer from what is called time-inconsistency. Using varying discount rates current optimal plans might not be consistent with what the same individual regards as optimal in the future (even in the absence of uncertainty about future states of the world and preferences). Hence, one does not stick to the original plan, *i.e.* policies are revised (if possible) as time passes and these deviations are (or at least could be) anticipated. If policy outcomes are irreversible, there might be regret. For example, although it seemed optimal to a social planner to convert some parcel of land, he/she later might regret this decision but be unable to restore its original state. While the occurrence of time-inconsistency under time-varying discount rates is widely accepted, there is a debate about whether this is actually a relevant problem (Pearce *et al.*, 2003).

Throughout, a key challenge in comparing costs and benefits of biodiversity policies across time by means of discounting is to answer the question of what constitutes the "right" discount rate. While this question arises with exponential as well as with hyperbolic discounting, with the latter it has an additional dimension. The problem is not only to pick a single parameter (which is difficult enough) but to choose an entire profile of discount rates. The use of hyperbolic discounting in the UK (Box 4.2) shows that important recent advances in addressing this problem can be reasonably implemented in real world policy. Transferring the approach to biodiversity policy would be straightforward.

4.3. Heterogeneous generations

Differences in discount rates used by individuals or groups within a generation are perhaps even more common and exaggerated than across generations. Box 4.2 shows the UK's discount rate starting at 3.5% in the early years, but then falling over time. The decline in the rate for UK policy is considerably smaller than the difference in discount rates across countries (Table 4.3). Rates vary by a factor of four, even within this relatively homogeneous sample. Less developed countries usually have higher discount factors as they rely on natural resources more than developed countries. Resource dependence is often associated with a higher discounting of future benefits since resources provide limited ability to distribute consumption over time. Poor countries that rely on consumptive aspects of biodiversity are more likely to manage their resources well, given that they rely on them for survival (though Diamond, 2005, provides some counter-examples). Nonetheless, subsistence is often used as a model of situations of high discount rates because the lack of sufficient reserves causes decisions to be made mainly on the basis of short-term considerations. Moreover, the non-consumptive

PEOPLE AND BIODIVERSITY POLICIES – ISBN 978-92-64-03431-0 – © OECD 2008

Box 4.2. Hyperbolic discounting in the *UK Green Book*

The evaluation of public policies in the UK is based on HM Treasury's (2003) *Green Book, Appraisal and Evaluation in Central Government*. For all projects with impacts lasting for less than 30 years a constant discount rate of 3.5% has to be used, based on empirical estimates for the UK. However, for policies with long-term effects the following pattern of discount rates is applied.

Table 4.2. **The declining long-term discount rate**

Period of years	Discount rate
0-30	3.5%
31-75	3.0%
76-125	2.5%
126-200	2.0%
201-300	1.5%
301+	1.0%

The effect of this stepwise decline in the discount rate is presented in Figure 4.2. While for the first 30 years the evolution of discount factors matches that of a constant 3.5% rate, for later periods the weight of future flows is significantly above that reference scenario, *e.g.* the weight put on any cost or benefit in year 200 is about six times as high with the declining discount rate than in the scenario with a constant rate of 3.5%; in year 300 the difference is already two orders of magnitude.

Figure 4.2. **Discount factors with decreasing discount rate of the *UK Green Book***

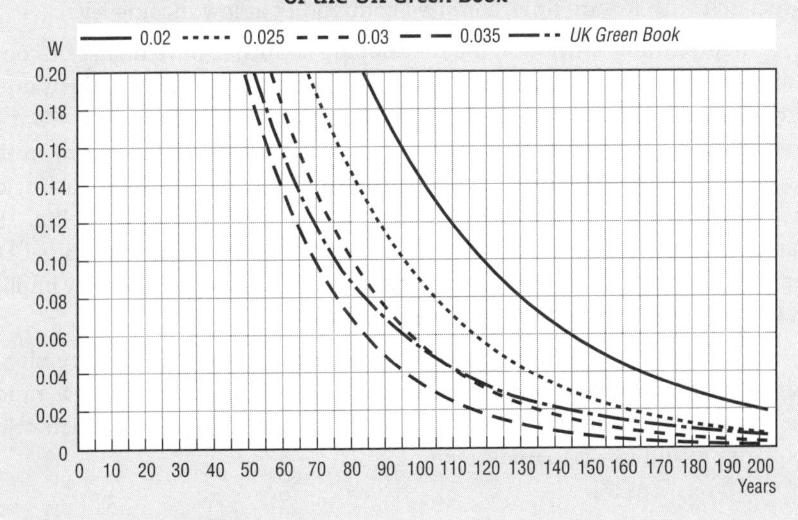

Source: HM Treasury (2003), Annex 6.

Table 4.3. **Discount rates as listed by Commissariat Général du Plan in France**

Country	Discount rate	Time horizon (years)
South Africa	8%	20-40
Germany	3%	Variable
Australia	6-7%	20-30
Canada	5-10%	20-50
Denmark	6-7%	30
United States	3-7%	Variable
Italy	5%	
France	8%	30
Hungary	6%	30
Japan	4%	40
Mexico	12%	30
Norway	5%	25
New Zealand	10%	25
Netherlands	4%	30
Portugal	3%	20-30
Czech Republic	7%	20-30
United Kingdom	3.5%	30
Sweden	4%	15-60
European Commission	5%	
World Bank	10-12%	

Source: Hepburn (2006).

benefits of biodiversity are likely to be highly discounted since they are associated with leisure time, a limited commodity at low income levels.

Underpinning the notion that developing countries have higher discount rates than developed countries is the observation that "liquidity constraints" force individuals to behave as if they cannot plan for the long term. That is, even when they know that postponing an action (*e.g.* consumption) to the future will bring greater overall benefit, they may be prevented from acting on that knowledge when they are unable to borrow against future benefits. The classic example is the farmer whose circumstances force him/her to eat the seed crop that was to be used to plant next season's crop. Such activity implies an extraordinarily high discounting of the future.

This heterogeneity across countries implies that when biodiversity is unevenly distributed globally, there will be differences in how much conservation one country is willing/able to undertake, and how much other countries would like it to undertake.

 PEOPLE AND BIODIVERSITY POLICIES – ISBN 978-92-64-03431-0 – © OECD 2008

4.3.1. *Intergenerational equity and intragenerational equity*

As most biodiversity policies have both long-term effects and affect people of different wealth we need to trade-off distributional effects against those across generations. For discussion on how to incorporate intragenerational equity into cost-benefit analysis see Section 2.3 of this book.

Helping the poor of today might harm future generations or *vice versa*. Hence, it is important to choose a consistent way to treat people living at different points in time and with different economic status. If the interests of the poor today are valued more than the interests of the rich, and if future generations tend to be better off than current ones, there is a case to be made for applying the same principle in both situations. The *Stern Review on the Economics of Climate Change* (Stern, 2006) has been criticised for being inconsistent on this important issue (*The Economist*, 2006).

The main concept that links the two issues is the elasticity of the marginal utility of consumption. This states by how many percent the utility of a person increases if his/her income increases by one percent. This is thought to be roughly constant across income classes. Hence, giving the same amount of money to a poor person (for whom it adds significantly to current consumption) creates more utility than if it is given to a better off person (for whom it is just a further drop in an already large pool). For consistency of the two dimensions of equity the same income elasticity has to be used when calculating the effective discount rate and distributional weights used to adjust for different income levels among members of the same generation. The *UK Green Book* (HM Treasury, 2003) assumes an elasticity of one in both cases.

4.4. Summary and conclusions

Part I has introduced the key concepts relevant to the analysis of distributive impacts of biodiversity policies – efficiency, cost effectiveness and distributive impacts – and how they relate to policies for maintaining and improving biologically diverse habitats and ecosystems. It has explained the role of CBA for biodiversity policy-making and how the integration of efficiency rules based on CBA has strengthened the case for biodiversity policies to be considered as important alongside other policy issues. At the same time, it has stressed that by using the concept of net social gains, CBA is severing the ties with richer welfare-theoretic dimensions.

We have explained how distributive impacts can be empirically measured, quantified in terms of summary values, and communicated in a policy-making context. We have also presented a positive analysis of the distributive effects of biodiversity policies, drawing from a rich set of case studies and examples to explain the links between policy objectives, instrument choice and welfare outcomes, while bearing in mind the success of the policy in maintaining and improving species-rich habitats and ecosystems. Important distributive dimensions that are dealt with are the spatial and the intertemporal distribution of welfare impacts associated with biodiversity-related policies. Our main conclusions are as follows:

- There are many suitable approaches for measuring distributive impacts, with differing data requirements and ability to capture these impacts. However, not all measures are equally useful in different conservation contexts. Hence, there is a need to develop criteria for judging the information contained in different measures and their adequacy for specific contexts.
- The distributive impacts of biodiversity policies are real and in many cases non-marginal. The primary distributive effects of biodiversity policies are likely to be pro-rich for both theoretical and empirical reasons. The secondary distributive effects are determined by the choice of instruments, which can mitigate or amplify the primary distributive effects. A wide variety of instruments is available for mitigating and potentially reversing distributive effects.
- The trend towards market-based instruments in biodiversity policies is likely to ameliorate the negative impacts on the poor of traditional instruments, such as protected area policies. However, there is evidence that while market-based instruments do not hurt lower income households, higher income households tend to profit relatively more.
- A "spatial mismatch of costs and benefits" (Wells, 1992) has been identified for some biodiversity policies, with local people often bearing most of the costs and populations of far-away countries receiving most of the benefits. However, if handled well, protection for conservation can be pro-poor because protected areas spell out – often for the first time – the precise nature of use rights in an area. Local communities may gain from protected areas because outside competitors are excluded.
- At the international level, current distributional problems are likely to persist. Many of the difficulties in translating the international willingness to pay for biodiversity conservation into flows of funds to areas of high conservation importance remain unresolved. The prevailing internationally agreed sharing rules for gains from international co-operation on biodiversity conservation contribute to this outcome.
- There is a significant intergenerational distributive dimension of biodiversity policies, since biodiversity policy decisions today will affect individuals currently alive, as well as generations not yet born. Ensuring that decisions taken today do not affect future generations can be addressed by varying the discount of costs and benefits arising at different

 PEOPLE AND BIODIVERSITY POLICIES – ISBN 978-92-64-03431-0 – © OECD 2008

points in time (hyperbolic discounting). At the same time, consistency between inter- and intragenerational equity is required.

With these key concepts, measurement methods, and empirical observations in mind, we now turn to the question of whether policy-makers should consider and address distributive issues *within* biodiversity policies – or whether these distributive issues should be left out of the picture when deciding between different courses of action.

PART II

Addressing Distributive Issues

ISBN 978-92-64-03431-0
People and Biodiversity Policies
Impacts, Issues and Strategies for Policy Action

Chapter 5

Should Biodiversity Policies Address Distributional Issues?

The starting point of this book is that biodiversity policies can create both winners and losers. This capacity is likely to be more pronounced the more the policy objectives deviate from the current *status quo* and the more uneven the distribution of income and wealth was prior to the implementation of the policy.

Part I has substantiated the relationship between socioeconomic status of individuals, households, groups, communities and countries and their reliance on biologically diverse ecosystems and habitats. It has also detailed how different instruments used to implement biodiversity policies have different distributive impacts. The key messages of Part I are that distributive impacts are real and pronounced, that historically there has been considerable use of those types of instruments that put a significant amount of the burden of conservation policies on poorer households, but that also reward-based instruments do not necessarily benefit poorer households more than the better off.

In this part, we consider to what extent these distributive issues should be made to matter when designing biodiversity policies. We first explain why economic welfare theory cautions policy-makers to include explicit distributive objectives within their policies. We then consider reasons why distributive issues are a relevant concern and what arguments policy-makers can use to support an explicit distributive dimension in terms of policy design, instrument choice and implementation.

5.1. Choosing between biodiversity policies when efficiency and distribution can be separated

5.1.1. *Assessing welfare impacts*

The aim of biodiversity conservation policies is to generate aggregate welfare gains to society. But in reality, policies usually lead to welfare gains for some and welfare losses for others.

Finding how to weigh these gains and losses against each other is fraught with difficulties. Welfare economists have developed various tests as a way of resolving them (Box 5.1). We discuss three prominent examples, namely the Pareto criterion, the Kaldor-Hicks compensation criterion, and the social welfare function approach (Just *et al.*, 2004). What these three approaches

 PEOPLE AND BIODIVERSITY POLICIES – ISBN 978-92-64-03431-0 – © OECD 2008

Box 5.1. **Tests of policy effects on welfare**

- Pareto efficiency is named after Vilfredo Pareto, an Italian economist. Given a set of alternative allocations of, say, goods or income for a set of individuals, a movement from one allocation to another that can make at least one individual better off without making any other individual worse off is called a Pareto improvement. An allocation is Pareto efficient or Pareto optimal when no further Pareto improvements can be made. Application of the Pareto criterion (PC) ensures that policies that are carried out will be unambiguously welfare improving and hence contribute to overall efficiency. Its shortcoming as a guide to policy-making is that it is highly restrictive: even policies that would contribute hugely to overall welfare are not admissible if a single individual incurs a loss, however small.
- The Kaldor-Hicks compensation criterion (KHCC) is a response to the restrictiveness of the Pareto criterion. It is a criterion of hypothetical compensation that requires the application of a test to any prospective policy: *Suppose those gaining from the policy could offer compensation to those who would lose from the policy. If there is an amount of compensation from gainers to losers that* i) *would make the losers voluntarily accept the policy and* ii) *would leave the winners better off with the policy than without, then the policy passes the test.* Three observations can be made about this test. One is that for a policy to pass the KHCC, it is not necessary for compensation actually to be paid. The potential is sufficient. The second is that if the payment of compensation is part and parcel of the policy, then the KHCC is equivalent to the PC. The third observation is that the KHCC is blind to distributive outcomes. Policies that exacerbate existing inequalities in income and wealth are still compatible with the KHCC. Refinements of the KHCC (such as the Scitovsky reversal criterion or Little's criterion) were developed in the 1950s (Just *et al.*, 2004; Nath, 1969), but the KHCC still stands, in particular in combination with cost-benefit analysis, as a benchmark test in public policy evaluation (Just *et al.*, 2004; Weimer and Vining, 1998).
- The social welfare function (SWF) involves a complete and consistent ranking of policies in terms of desirability (a social welfare ordering). If such a social welfare ordering is continuous, then it can be represented by a SWF which evaluates and aggregates the utility levels of all individuals in society resulting from the policy into a single value. The problem with the SWF is specifying a function that fulfils certain axiomatic requirements that a reasonable person might want to impose on it. Arrow (1950) has shown that such a SWF cannot be constructed. However, relaxing one or more of these requirements allows distributive issues to be explicitly considered in this framework (Sen, 1997). The concept of the SWF itself provides no guidance on how inequality should be considered. We comment on this problem below.

have in common is that they are invariably grounded in a utilitarian approach to welfare measurement.

5.1.2. *Implications for policy choice*

When combined with cost-benefit analysis, the KHCC becomes a powerful tool for choosing between policies. CBA allows the evaluation of policies using money equivalents of aggregate social welfare gains. In other words, CBA helps policy-makers arrive at a dollar or Euro estimate of the gains delivered by different policies. These gains are the sum of social benefits of a policy minus the sum of social costs. Within the limitations of CBA (see Serret and Johnstone, 2006), the KHCC allows different policy options to be ranked by the volume of net gains created, but with no regard to its distributive impacts. This ranking then provides an unambiguous foundation for determining the right policy.

Choosing a biodiversity policy on the basis of the KHCC has clear advantages:

- It leaves biodiversity policy unconstrained so that the most efficient means of bringing about the desired environmental outcome can be pursued.
- It retains the focus of the policy on the allocation problem at hand and thus reduces the complexity of the decision-making process.
- It recognises that different socio-economic groups will be differently affected on the basis of their income, asset base, preferences, employment opportunities, etc.
- But it leaves the implementation of a "fair" distribution of income and wealth to the domain of redistributive policy-making (*e.g.* through taxation), and it requires a specific political legitimacy to carry out such measures.

In essence then, using the KHCC allows policy-makers to separate efficiency considerations, important for choosing between biodiversity policies, from equity considerations, which arise out of the distributive effects of the implementation of such policies. Whatever distribution of income and wealth is regarded as desirable can then be implemented in the most efficient manner, so as to minimise the equity-efficiency trade-off. Fiscal authorities can therefore concentrate on redistribution measures involving the least deadweight loss of taxation, starting from lump-sum transfers between different households.

Removing the distributive aspects from biodiversity policies has therefore much to recommend it: biodiversity policies do not need to be concerned with ensuring desirable socioeconomic outcomes that may only be accomplished at great cost to the biodiversity policy's efficiency. Any

 PEOPLE AND BIODIVERSITY POLICIES – ISBN 978-92-64-03431-0 – © OECD 2008

undesirable redistributive effects can be managed in a second round in the most efficient manner, thus ensuring that the net gains of the biodiversity policies are realised to the greatest possible extent.

Looking at actual policies, there is evidence that current instruments used to accomplish distributional objectives within biodiversity policies frequently lead to efficiency losses and unintended distributional consequences.

Efficiency losses

Efficiency losses are most palpable in policies that explicitly aim to be equitable. In many situations, policy-makers attempt to make policies more acceptable by choosing instruments that "treat everybody the same". In practice, this commonly translates not so much into equitable policies, but into uniform policies, the advantage being that uniformity is much easier to communicate and verify than equity. Typical and representative examples are uniform compensation payments and uniform regulations. Under uniform payments, every landowner receives an identical per-area payment for participating in a policy, irrespective of the individual costs of compliance and of the marginal benefits generated on different sites. Under uniform regulations, landowners typically have to change management practices on the same amount of land or on the same share of landholdings, again irrespective of underlying heterogeneities. A moment's reflection is enough to see that such uniform policies can be efficient only when every policy participant is the same. When landowners differ in terms of the marginal benefits generated per unit of land dedicated to conservation, there would be gains from landowners being treated differently. Box 5.2 summarises studies into conservation contracting in Germany, which calculate considerable waste of resources from uniform payments. In this case, cost savings of up to 70% could be achieved by moving away from uniform payments.

Other examples of distributional objectives are policy instruments that aim to improve conservation *indirectly* (see Ferraro and Simpson, 2002). Typical examples of such indirect instruments are price guarantees for agricultural products grown in accordance with biodiversity protection guidelines (*e.g.* shade-grown coffee) and promotion of ecotourism which requires that local ecosystems are protected by local communities.

What efficiency losses are inherent in indirect policies? While these instruments have correctly been argued to both improve conservation and deliver benefits to locals participating in the conservation effort (Pearce and Moran, 1994; Heal, 1999), they have not been demonstrated to be the most cost-effective means of doing so. Ferraro and Simpson (2002) showed that direct policies (*i.e.* payments to individuals and groups for the delivery of

Box 5.2. **Contracted conservation in Germany**

Many endangered plant and animal species depend on the extensive use of agricultural land that has shaped the landscape of large parts of Western Europe in the past. Modern, more efficient agricultural practices threaten their survival by habitat conversion. Direct contracts between a state agency and (mostly) local farmers form the backbone of Europe's conservation efforts. They are designed to create incentives for farmers to continue to manage at least some of their land in a traditional way. Hence, farmers are compensated for not exploiting the full economic potential of the land in order to conserve specific characteristics that are important for target species and habitats. Between 2000 and 2005, the EU spent about EUR 2 billion per year on such payments for conservation service schemes. This is topped-up by contributions from individual member states, since the EU's share amounts to only 60-85% of total costs. On average about 50% of member states' budgets for rural development is for this type of incentive scheme. This results in about 25% of all agricultural land being subject to voluntary conservation restrictions (European Commission, 2005).

Participation in the German conservation contract scheme (*Vertragsnaturschutz*) is voluntary and is sometimes limited to a certain percentage of the total cultivated land of each farm (*e.g.* 30% in the state of Schleswig-Holstein), in order to prevent economic dependence on the scheme. Payments per hectare are fixed and depend on the measures taken. Hence, spatial heterogeneity in benefits due to geographical variations or due to the contiguity of preserved habitats is not reflected in the level of compensation. Two arguments in favour of this uniform payment have been made: a) tailor-made schemes involve excessive transaction costs; and b) being equitable is crucial for the acceptance of this programme by potential participants (Ohl *et al.*, 2006).

However, homogeneous payments come at a cost. Wätzold and Drechsler (2005) showed that spatially uniform payment schemes are inherently inefficient when marginal benefits differ across sites. Given equal marginal costs, a site with higher marginal benefits should receive higher payments than a less valuable one. Otherwise, too much land of low ecological quality is conserved, eating into financial resources for more valuable sites. A reallocation of payments could therefore increase conservation impact without affecting the budget. When agglomeration of conserved areas is relevant for ecological effectiveness, as Drechsler *et al.* (2007) showed for a scheme in Germany to protect the scarce large blue butterfly (*Maculinea teleius*), a heterogeneous payment mechanism could save up to 70% of costs compared to a uniform scheme.

habitat protection and ecosystem maintenance) will usually be more efficient than indirect policies. The reason is that indirect instruments rely on more complex pathways to accomplish their objectives, leading to distortions (in the form of spill-overs into other markets), friction (in the form of transaction costs along the causal chain) and slippage (in the form of better informed participants extracting information rents). Also, direct instruments generally have simpler institutional arrangements, hence lower costs of administering the scheme (Ferraro and Simpson, 2002). This means that either more conservation could have occurred for a given level of funding, or that the given level of conservation could have been achieved at lower cost.

Unintended distributional consequences

The unintended distributional consequences of indirect policies are a result of their more complex causal pathways, as mentioned above. A typical example is the funds generated by consumers paying price premiums for organically produced food or "fair-trade" produce. In many of these cases, consumers perceive that buying these products helps reconcile the biodiversity benefits of low impact agriculture in developing countries with the cost to local producers of foregoing the benefits of intensification. These consumers are characterised by a surprisingly low price elasticity of demand (Arnot *et al.*, 2006). However, there is evidence that in some important cases, only a small fraction of the funds raised through price premiums reach the local level. The reason is that other parties along the chain may capture a greater share of the price premium than the local producers. A recent estimate for Fairtrade coffee sold in coffee bars put the share of the premium going to producers at only 10%, with the largest share accruing to retailers (Harford, 2003).

To sum up, separating equity and efficiency objectives liberates biodiversity policy-makers from additional constraints and obligations, leaving them free to pursue those policy options that promise to deliver the greatest social gains. This would imply, in simplified terms, that biodiversity policy should therefore be chosen by using CBA to select the highest ranked policy that is feasible.

However, separating efficiency and equity in biodiversity policies is difficult for several reasons: *i)* the strong public goods aspects of biodiversity, the informational imperfections inherent in biodiversity policies, and the transaction costs in carrying out redistributive transfers after the policy is implemented; *ii)* practical limitations to separability, such as obstacles to implementing the redistributive part of the separable policy. When distributive effects of biodiversity policies are non-marginal and redistributive policies cannot be relied upon to re-establish a desirable distribution of income and wealth following implementation of the biodiversity policy, then biodiversity policies have to assume more of the weight of integrating equity concerns into the policy itself. We discuss these points in more detail in the next section.

5.2. Challenges in separating efficiency from distribution

In applying the Kaldor-Hicks criterion, it is assumed that policy-makers can implement redistributive measures alongside any specific policy. In particular, the expectation is that such measures would restore the pre-policy wealth distribution. The criterion does not govern whether or how individuals or communities that lose out as a result of the policy are compensated. This truncation of the analysis before any actual compensation is paid is commonly justified by the "statistical evening-out" effect: different individuals will incur small losses while there will be significant social gains at the aggregate level. The policy with the highest net surplus should be selected. However, in the following section we discuss a number of reasons why the simple rule of maximum net surplus may not be applicable.

5.2.1. *Public goods aspects of biodiversity*

A distinctive aspect of biodiversity is the "public good" nature of many of the goods and services it provides. This limits the extent to which efficiency and distribution can be separated, because public goods, by definition, are consumed by everyone in the same quantity.

This public goods property gives rise to an explicit efficiency-equity link. Chichilnisky and Heal (1994) showed this for another environmental global good: carbon emissions. A common recommendation to ensure efficiency in international greenhouse gas abatement is to require that marginal costs of abatement be equalised across all emitting countries. This rule, however, has non-marginal distributive effects. Extra income typically provides lower welfare gains for those on high income as opposed to those on low income (diminishing marginal utility of income). It is therefore reasonable to assume that the marginal utility of consumption is highest in developing countries. Thus, when the marginal cost of abatement of carbon is equal everywhere, the marginal loss of utility will be disproportionately high in developing countries. This clearly implies a strong re-distributive impact from such an abatement policy. In fact, for a policy implementing this efficiency rule to pass the PC, it would have to start by treating countries differently: it would need to attach lower weights to the marginal utility of consumption in countries with lower incomes. In other words, the only way – from a distributional perspective – to justify the policy commonly advocated for carbon emissions is to minimise the accounting of the loss incurred by developing countries. Indeed, attaching equal weights to the marginal utility of consumption would result in an

PEOPLE AND BIODIVERSITY POLICIES – ISBN 978-92-64-03431-0 – © OECD 2008

efficiency rule that requires a higher contribution from countries with higher incomes. This demonstrates that since countries are inexorably linked together through the global public goods dimension of the environmental problem, efficiency and equity are no longer separable and some policy issues may need to be reconsidered.

For biodiversity policies that generate international public goods, Chichilnisky and Heal's (1994) findings provide a sound economic basis for arguing that – on efficiency grounds alone – higher income countries should contribute proportionately more to the cost of preserving habitats and ecosystems than poor countries. This argument gains weight when one considers that demand for the public goods aspects of biodiversity increases with income, as with almost everything consumed.* A well-developed tradition in public finance says that when people of similar preferences have different incomes, those with higher incomes pay disproportionately more for public goods. This has long been used to justify progressive taxation on efficiency grounds (Musgrave, 1959).

Chichilnisky and Heal's analysis therefore favours biodiversity policy payments from richer to poorer countries. It also offers an alternative to the idea of incremental cost as the basis for international transfers. While incremental cost is based on the compensation of costs incurred for the marginal unit of biodiversity provided, transfers in the spirit of Chichilnisky and Heal (1994) would be based on a measure of benefits derived. The argument for such a biodiversity-specific transfer is incomplete, however, without showing that such policy-specific payments are superior to simple lump sum transfers between countries. After all, it could be argued that unspecific transfers from high-income to low-income countries, for instance in the form of development aid or assistance by international lenders, already compensate low-income countries. The argument for policy-specific payments in proportion to income therefore needs to demonstrate that unspecific transfers are less effective than policy-specific payments.

There is a large literature in public economics discussing the merits of specific *versus* unspecific transfers, starting with the long discussion on the idea of tax hypothecation ("earmarking"). Applied to the context of biodiversity, there are important arguments in favour of specific payment:

- Biodiversity-specific transfers would limit the possibilities for rent appropriation by the budgetary authority, both when raised in the donor country and spent in the receiver country (Buchanan, 1963).

* *E.g.*, instead of always eating home-cooked meals, people with higher income eat prepared foods and more expensive restaurant meals, so money spent on a broadly defined "food" keeps pace with income.

- International biodiversity-specific transfers are more robust to changes in government and administering institutions (Brett and Keen, 2000).
- The biodiversity-specific earmarking of transfers would provide a commitment device if any one of the governments faces a credibility problem regarding the policy areas on which public funds are spent (Marsiliani and Rennström, 2000).

In sum, political support for such policy-specific payments would be higher since they appear to restrict their use and require less trust in politicians.

5.2.2. *Efficiency costs of redistribution*

Even if compensation is carried out after policy implementation, economists point to important reasons why efficiency and equity may still be closely related. These reasons are inherent in the "leaky buckets" (Okun, 1975) in which income is redistributed from winners to losers.

The leaky bucket argument is at the heart of the literature on the efficiency-equity trade-off in policy-making. Its main message is that any redistributive policy is inefficient because of two distortions:

1. The transaction costs of moving a dollar from one person to another. Since this move requires resources in the form of fiscal assessment, administration etc., out of every dollar raised from contributors at one end of the transaction, less than one dollar reaches the recipient.
2. The incentive effects of redistribution (Mirrlees, 1979). Most redistributive transfers are not lump sums, but tied to work effort or other productive choices of the contributor. As taxation can deter resources from being efficiently employed by the potential contributor, raising a dollar of money to be transferred costs more than one dollar in terms of output foregone. Empirically, this argument is tied to estimates of the so-called "marginal social cost of public funds". Recent estimates for the United States put this cost at between USD 1.28 and USD 1.70 for every dollar raised (Allgood and Snow, 1998).

Some observers use the presence of these distortions to argue for as little redistribution to be carried out as possible (Okun, 1975). Alternatively, their presence raises the possibility that redistribution, if desired, can be carried out more efficiently within biodiversity policies rather than being left to an imperfect redistribution system after the policy has been implemented. If biodiversity policies can achieve redistributive objectives cheaply, *i.e.* with little loss in efficiency, they should be considered. We discuss candidates for such policies in Part III.

5.2.3. *Information*

The third fundamental problem with separating efficiency and distribution in biodiversity policies is the fact that not all parties concerned are well informed about the nature of the policy and its impacts. This concerns at least two parties in the "linear policy-making model":

- The groups of individuals who are immediately affected by the implementation of the biodiversity policy.
- The policy-makers involved in determining the policy objectives and instruments.

It is well known that imperfect information among either of these parties can lead to significant distributional effects (Boyce, 2002). If those affected by the policies are badly informed about functional linkages, externalities and other dimensions of the biodiversity problem, this will prevent these groups from: *i)* expressing their true preferences through their activities; and *ii)* adequately expressing their true preferences in the political process that gives rise to policy interventions. Since access to information and the ability to use the information productively will be lower for poorer households, it is likely that providing information would be the cheapest way of correcting the inefficiency (Serret and Johnstone, 2006).

5.3. Practical limitations to separating efficiency and distribution impacts

While the arguments above seek to point out the fundamental reasons for considering the efficiency and the distributive impacts of biodiversity policies together, there are also a number of practical reasons limiting the separation of the two aspects.

5.3.1. *Spatial and functional scope of jurisdictions*

For compensation to be paid, coercive transfers (such as taxes) have to be feasible. This is only possible if there is a geographical overlap between the winners and losers and the scope of the institution making the transfers. All countries have institutions at the local, regional and national level that can – in theory – bring about transfers between households. These institutions are state bodies with fiscal authority and are commonly charged with carrying out the redistributive part of government policy.

At the international level, however, institutions with coercive fiscal authority do not exist. No mechanism exists therefore for compulsory taxation of beneficiaries outside the boundaries of those jurisdictions that are implementing the policy. The only exception is regional associations, such as the European Union. However, even in these cases, the coercive power to "tax" countries is heavily circumscribed.

So for biodiversity policies that generate global benefits to those beyond the relevant jurisdictions, transfers can only be made on a voluntary, rather than on a coercive basis. This severely restricts the ability of policies to raise the necessary funds for compensating for the non-marginal distributive effects of biodiversity policies.

There are also functional limitations to redistributions between winners and losers. In many countries with low levels of economic development the relevant institutions have severe limitations to carrying out their redistributive functions (Schneider, 2005). Recent estimates put the volume of tax revenue lost by developing countries as a result of the functional limitations of their tax collecting institutions at more than the total volume of direct external development aid (Cobham, 2007). Additionally, since households with higher incomes benefit disproportionately from tax evasion and tax avoidance, these functional limitations have a strongly regressive distributive impact.

5.3.2. *Common property resources*

Habitats and ecosystems are frequently managed in the context of common property institutions and practices. In such settings, both the degree of efficiency of aggregate resource management decisions and the institutions that co-ordinate individual management decisions are endogenous to the distribution of income and wealth among those participating in the management. For example, the production efficiency of rural co-operatives may benefit in particular from the participation of a single well-off farmer since only he is able to afford specific types of human or physical capital (Datta and Kapoor 1996). In such settings, (in)equity and efficiency can be inseparable; outside interventions that change the distributional patterns can lead to unpredictable changes in resource management. For example, when individuals whose contribution is crucial withdraw from the CPR system, equity may be enhanced, but efficiency suffers disproportionately.

Both the theoretical and empirical literature on common property resources (CPR) stresses the causal links between efficiency (in terms of successful management) and distribution. The management of common property resources can be framed as private provision of a public good. The public good of concern in CPR is resource conservation. Poor management of an open access resource due to the absence of hierarchically superior institutions is called the "tragedy of the commons" (Hardin, 1968). Consumption of the common resource by one agent restricts use by others. If these negative externalities are not taken into account, the resource is

 PEOPLE AND BIODIVERSITY POLICIES – ISBN 978-92-64-03431-0 – © OECD 2008

overused and the allocation is inefficient. Warr (1983) pointed out that in a setting in which individuals privately contribute to a public good, the size of the inefficiency is independent of the distribution of income among users. This is the basic rationale for separating efficiency and distribution in CPR. Bergstrom *et al.* (1986) suggested that if redistribution changes the number or identity of users, aggregate resource consumption and, hence, efficiency, are affected.

Both Warr (1983) and Bergstrom *et al.* (1986) derive their results from stylised settings in which all users are equally affected by excessive resource use and conservation of the resource does not require any fixed upfront investments such as a switch of technology or institutional setting. However, neither condition is commonly met in reality. Baland and Platteau (1997) showed that if users differ in the extent to which they are affected by the quality of the resource, an increase in inequality among the potential users of the resource can have significant effects on efficiency. A small redistribution from rich to poor can induce a collapse of conservation efforts. Taking income or access rights from the better off can reduce their willingness to contribute to conservation in a way that cannot be compensated for by the increase in the willingness to contribute by the poor who gain from redistribution. This is due to a co-ordination problem that arises when the provision of the public good involves non-convexities (*e.g.* fixed costs). Unless large contributions by the rich pass a certain threshold, contributions from all users collapse. Hence, a narrow focus on the poorest users of a CPR can result in disastrous conservation policies.

Another way in which inequality can influence efficiency is if users can change their ability to extract from the resource, *e.g.* by investing in machinery. Aggarwal and Narayan (2004) found that users invest in excess capacity except when inequality is very high. In these cases efficiency of resource use is highest when inequality is either very low or very high.

A number of empirical studies estimated the impact of different proxies for household wealth (*e.g.* education, land holdings, etc.) on resource use for different CPRs. As Table 5.1 shows, the results are mixed. While education seems to decrease households' reliance on income from CPRs, the effect of other assets and alternative sources of income is less clear. In some cases reliance increases with economic status, while in others the opposite is true.

The initial distribution of wealth, skills and access rights does also affect the performance of a policy imposed externally on a commonly managed resource. Baland and Platteau (1998) found that such interventions tend to increase inequality among users. The poor are therefore more likely to be hurt by such measures. Moreover, if a policy is required to benefit all users, *i.e.* to be Pareto superior to the original state of the world, its efficiency gains decrease

Table 5.1. **Empirical estimates of the relationship between distribution of wealth and resource use in CPRs**

Authors (year)	Resource	Region	Inequality measure	Impact on resource use or dependence
Adhikari *et al.* (2004)	Forest	Nepal	Land and livestock holding Education Economic status	+ – +
Adhikari (2005)	Forest	Nepal	Economic status Livestock Education	+ + -
Barbier/Cox (2004)	Mangrove forest	Thailand	Economic status (of region not household)	–
Fisher (2004)	Forest	Malawi	Economic status	–
Fisher *et al.* (2005)	Forest	Malawi	Economic status Education	– –
Quang/Anh (2007)	Forest	Viet Nam	Access to other resources Female labour share	- +

as the inequality among users increases. This holds for commonly-used instruments such as quotas and taxes applied uniformly.

The empirical evidence on the performance of CPR management is mixed. While the "tragedy of the commons" is a real phenomenon, it does not imply that all commons are managed badly by their local community of users (Ostrom and Gardner, 1993; Hegan *et al.*, 2003). However, successful schemes tend to collapse when there are rapid changes in the population of users, technology, and economic and social conditions (Dietz *et al.*, 2003).

Recent empirical evidence affecting environmental and distributive impact of regulatory schemes in CPRs is summarised in Table 5.2. Only one study found a progressive impact of the conservation policy. This strongly supports the theory that conservationist interventions tend to favour the better-off users of a CPR.

Table 5.2. **Examples of impacts of policies regulating CPRs**

Authors (year)	Resource	Region	Instrument	Env. impact	Distributive assessment
Beukering *et al.* (2003)	Rainforest	Indonesia	Park (Leuser National Park)	+	Progressive
Chakraborty (2001)	Forest	Nepal	Privatisation	+	Regressive
Lybbert *et al.* (2002)	Argan oil	Morocco	Marketable use rights	n.a.	Regressive
Zbinden/Lee (2005)	Forest	Costa Rica	Payments for Environmental Services	n.a.	Regressive

n.a.: not assessed.

PEOPLE AND BIODIVERSITY POLICIES – ISBN 978-92-64-03431-0 – © OECD 2008

Equity and efficiency are inherently linked in the management of CPRs. Save for some very specific cases, the distribution of access rights, income and access to alternative sources of income have direct and significant effects on efficiency. Moreover, attempts to intervene in such schemes cannot untie this link. Both their effect on efficiency and equity depend on the initial distribution.

5.3.3. *Intergenerational compensation and its limits*

Part I of this book highlighted how distribution impacts of biodiversity policies vary over time, meaning that different generations experience different costs and benefits of the policy. These intertemporal aspects are also the third key practical limitation to separating efficiency and equity considerations. This is because there are few functional mechanisms for transfers across generations.

In theory, it is possible to think of financial mechanisms for transfers to be made between present and future generations. These mechanisms could either: *i)* compensate current generations for the costs of preserving ecosystems whose benefits will accrue to future generations; or *ii)* compensate future generations for habitat conversion or ecosystem degradation carried out by the current generation. In a classic paper, Gale (1973) described a simple framework in which the first generation can establish an intergenerational "fund" (through coercive taxation of the first generation) that will increase welfare for all subsequent generations. Bovenberg and Heijdra (1998) extended this model to an economy with an environmental sector in which public debt, passed on to future generations, can compensate the current generation for incurring costs of environmental protection to benefit future generations. For public debt to play this role requires sophisticated planning, both for pricing the environmental resource correctly and for issuing just the right amount of bonds in order for efficiency to be maintained. Most recently, building on Stern's (1997) idea of an "intergenerational democracy", Gerlagh and Keyzer (2001) defined the rules for an intergenerational trust fund as a redistribution mechanism. These specify that each generation has a fiscal entitlement from the trust that will allow it to purchase a minimum level of environmental goods and services.

As Howarth (2000) pointed out for climate change, these ideas are sound and conceptually appealing, but fraught with practical difficulties. As the literature on cost-benefit analysis makes clear, the problem of intergenerational mechanisms is hampered by three key problems (Lind, 1995):

- The imperfections of the political process: in democratic systems, subsequent policy-makers may not feel bound by decisions of previous governments about how to allocate public funds.
- The problem of commitment: even if subsequent generations of decision-makers do feel bound by decisions of previous governments, circumstances can change, short-run budget shortfalls may have to be covered, and decisions can be reviewed.
- Time consistency: even if an intergenerational transfer was optimal at some previous point in time, after its establishment it may no longer be optimal to see the transfer through to the end.

If intergenerational financial mechanisms are difficult to achieve, biodiversity policies may have to incorporate features that explicitly address these intergenerational costs and benefits. This is because the mechanisms managing intergenerational lump sum transfers cannot be relied upon to correct the non-marginal effects across longer time horizons.

5.3.4. Political economy factors

Political power

The fourth angle of attack on the idea of separating efficiency from equity is the presence of important differences in political power across different groups (Drazen, 2001; Bueno de Mesquita *et al.*, 2003). This was applied to the context of biodiversity by Deacon (2006).

The linear biodiversity policy model that gives rise to the logic that efficiency and equity should be dealt with separately suffers from a severe shortcoming: by leaving out the role of political power, it overlooks a key factor that shapes the choice of policy objectives and the choice of instruments. This is because these choices are not made mechanically, but within institutions of public choice. Heterogeneity in access to these institutions and ability to influence their processes and outcomes leads to what Deacon (2006) called the "traditional incidence pattern", whereby the poor are net losers from policy changes.

The argument mirrors Drazen's (2001) seminal study on how politics shape economic outcomes. It observed that prior to the introduction of biodiversity policies, groups are differently positioned to influence what policies are adopted. Some groups have *power*, with "power" meaning "the ability of an individual or group to achieve outcomes which reflect his objectives" (Drazen, 2001). The policy choice that results from the process therefore already contains strong redistributive components right from the start. Since there is a robust and fundamentally positive correlation between economic status and access to the institutions of public choice, this leads to several closely related consequences:

- Policies will reflect and reinforce prevailing patterns of political power and economic status (Drazen, 2001).
- The average policy will deliver the greater share of benefits to the better off households and the greater share of the costs to the poorer households (Deacon, 2006).

A good illustration is found in a counter-example to Demsetz (1967). Harold Demsetz argued that property rights are human institutions that are created and evolve with the needs of societies. This argument predicts that institutional changes will come about in response to changes in their costs and benefits. Libecap and Smith (2001) tested that prediction for the evolution of property rights in oil and gas extraction in the United States. They found that a number of factors, including political power, influenced the evolution and distribution of those rights in ways that could not have been predicted by the simple application of Demsetz's hypothesis. This illustrates the fundamental importance of political economy in policy outcomes.

This leads to some general hypotheses about distributive impacts of biodiversity policies in the presence of unevenly-distributed political power:

- Socio-economically, the costs of protected area policies will be shouldered predominantly by poorer groups.
- Spatially, this means a disproportionate concentration of conservation areas on lands used by poorer groups compared with lands used by groups of higher socio-economic status.
- Functionally, the presence of concentrated political power will lead to increased corruption in conservation policies in order to steer the benefits to favoured groups and to a reduction in the effectiveness of biodiversity policies (Deacon, 2006).

These economic arguments are empirically well substantiated. Part I of this book has documented how the greatest net benefits tend to accrue to the better-off. This was true for both voluntary and involuntary instruments, although the imbalance tended to be more extreme for involuntary instruments. Witness also the predominant use of coercive protected areas on lands previously used by poorer groups, with little or no compensation. These findings are widely echoed by observers in the field: "While state conservation policy shapes how national parks impact upon local resource access and use, older political economic inequalities [...] build on such policies to influence how management affects the livelihoods of poor households" (Dressler, 2006).

The functional link between the uneven distribution of political power, corruption and the effectiveness of conservation policies has also attracted empirical attention. In a recent report, the World Bank (2006) cited a document prepared for the American Forest and Paper Association that outlines the relationship between suspicious logging activities and levels of corruption in roundwood producing countries (see Figure 5.1).

Figure 5.1. **Corruption and illegal forest activity**

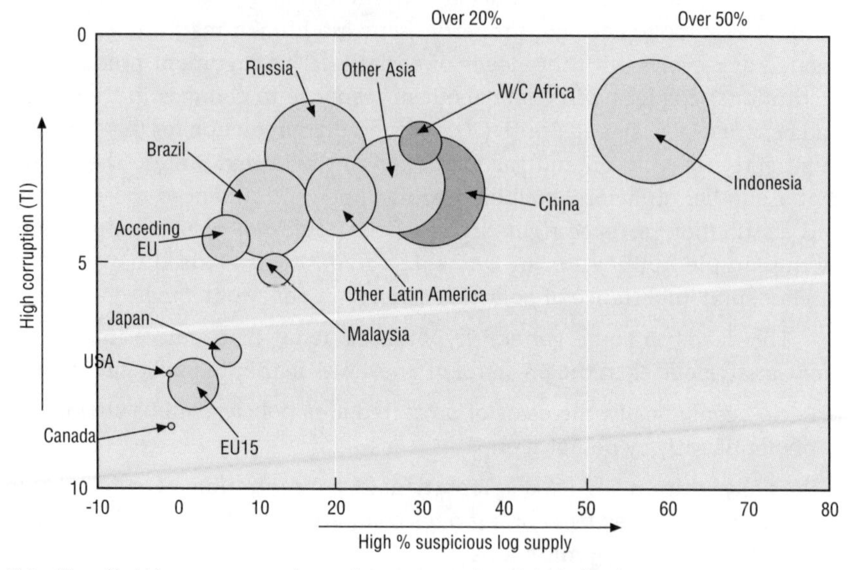

Note: Size of bubbles represents volume of suspect roundwood, including imports.

Sources: Transparency International; WRI/SCA estimates of illegal logging, cited in World Bank, 2006.

From the point of view of global efficiency in biodiversity policies, the costs of neglecting political power and corruption in policy choice are manifold (Kishor and Damania, 2006). Apart from the reduced effectiveness of the policy chosen on narrow conservation grounds, corruption:

- reduces the public tax and royalty base;
- inflicts economic damages on compliant groups;
- undermines the credibility of governance institutions; and
- damages the livelihoods of communities residing in affected areas (Kishor and Damania, 2006).

The extent to which different policies and their implementing instruments reflect entrenched interests and the extent to which they enable political power to be exercised through the policy chosen can therefore not easily be separated from the policy itself.

PEOPLE AND BIODIVERSITY POLICIES – ISBN 978-92-64-03431-0 – © OECD 2008

The political economy of conservation organisations

The emphasis on political power as a source of problems for the design and implementation of equitable biodiversity policies is primarily a dimension of income and wealth. In the context of biodiversity policies, however, there are other dimensions of political economy that are of significance. One key dimension is the political power of groups with a conservation objective (*versus* those with strong economic growth objectives). This dimension has taken on special significance in the wake of severe criticism of the large conservation organisations such as the Worldwide Fund for Nature (WWF), Conservation International (CI), and The Nature Conservancy (TNC) for neglecting the interests of traditional and indigenous peoples in the pursuit of conservation policies (Chapin, 2004). This criticism coincides with three developments over the last 10 years:

1. The decreasing overall volume of funds available for conservation projects in developing countries. Precise figures are difficult to compile, but one recent estimate puts the decline in annual expenditures from international sources for protected areas from USD 700-770 million in the mid-1990s to USD 350-420 million in 2004 (Khare and Bray, 2004).
2. The increasing concentration of conservation funds in the hands of a very few non-governmental organisations. Again, figures should be taken as indicative only, but some estimates put the so-called "Big Three" (WWF, CI, TNC) in control of about 50% of total available funding for conservation in 2002 (Khare and Bray, 2004).
3. The diversification of the funding base of the Big Three from an initial reliance on private funds to an increasing share of public funds from bilateral and multilateral donor organisations, as well as corporate funds (Chapin, 2004).

These developments raise two separate, but deeply related, issues in the context of the political economy of biodiversity policies. One is that many conservation policies are carried out in developing countries with inherently weak governmental institutions. In such a setting, it is possible for well-funded and well organised non-governmental organisations to find it easier to carry out policies whose distributive effects have not been subjected to sufficient scrutiny. This would point to a regulatory failure by weak governments. It is therefore conceivable that biodiversity policies could be implemented to a disproportionate extent in a manner that harms the interests and aspirations of indigenous communities.

The second issue is that in order to efficiently use biodiversity conservation funds from public donors, outsourcing implementation of biodiversity policies to specialised institutions is preferable to implementing the policies themselves. Such specialised institutions have sufficient on-the-ground experience and capacity to manage conservation funds effectively. The conservation organisations such as the Big Three have a comparative advantage in providing these services, fulfilling both the conservation objectives and the interests of the public at large. The challenge is that outsourcing policy implementation to conservation organisations gives rise to what economists refer to as "agency problems": the inherent objectives of the public donors and those of the conservation organisation may not be exactly the same. Specifically, public donors may attach a different weight to the distributive effects of biodiversity policies than the conservation organisation. This would not matter if the donor and the organisation could easily sign a binding contract governing the types of distributive effects to be brought about by the organisation; and if the donor could monitor the organisation's performance against the contract at little or no cost. In reality, however, contracting for distributional outcomes is difficult and monitoring is invariably costly to carry out. This leaves conservation organisations with greater control over how policies are carried out.

Thus, the political economy of conservation organisations means that the organisation of biodiversity policies for efficiency purposes and their distributive consequences cannot be fully separated. In the real world, given weak governmental institutions and agency problems, conservation organisations have additional scope for influencing policy implementation, in particular in developing countries.

Conservation organisations and government agencies

Easterbrook (2003) notes an important cultural difference between Europeans and Americans that has some implications for financing biodiversity in developing countries. Europeans and Americans are equally generous when looked at from a national perspective, but differ in how that generosity is implemented. Americans tend to make more private charitable donations than Europeans, but European governments allocate more of their budgets to charitable objectives. This is especially true in international aid, where there are significant differences in the composition of public *versus* private aid (as a proportion of national income). Comparisons that focus only on government aid (otherwise known as official development assistance, ODA), or only on private donations, will give a distorted picture.

For financing biodiversity, this has important consequences. The objectives of government agencies can often be different from those of private groups. The kinds of problems outlined by Chapin (2004) are specific to private groups and are generally less common to ODA, where government-to-government interactions more often mitigate such impacts. Private groups must necessarily engage in activities that enhance their ability to obtain additional funding – they operate in an environment of competitive

PEOPLE AND BIODIVERSITY POLICIES – ISBN 978-92-64-03431-0 – © OECD 2008

fundraising. As such, projects whose benefits might be long term, with little immediate impact, may be less likely to be undertaken with private money. ODA, on the other hand, does not have as strong a requirement to show immediate returns.

This distinction in outcomes, which is generated purely on the basis of the source of funding, is another reason for the difficulty in separating efficiency from distributive impacts. Consider a public policy agenda that may have two similar means of attaining its objectives. When the outcome is affected simply by the source of funding, then achieving efficiency in the policy is left to chance! This makes it difficult to argue that there is a separation between efficiency and distributive impacts since the distributive impacts themselves are linked to the type of funding that underpins the policy agenda.

5.3.5. *Conflict*

The fifth obstacle to removing redistributive concerns from biodiversity policies is policy-induced conflict. Policies involving strong redistributive effects, particularly at the expense of a well defined group of individuals, can mobilise opposition to the implementation of the policy, the subsequent management regime or both. This raises the cost of policy implementation and can make the pursuit of the original policy thoroughly unattractive on efficiency grounds alone (Bardhan, 1996). These costs arise in two forms: *i)* policy implementation has to be accompanied by costly visible and credible enforcement activities; *ii)* if parties find it worthwhile to challenge the policy through open conflict, then these costs subtract from the net gain generated by the policy implemented. The possibility of distribution-induced conflict alone is therefore relevant for policy choice.

Enforcement activities

Enforcement activities are costly, but necessary, for policies likely to induce conflict. This introduces a trade-off between effectiveness of the policy and the need for an enforcement budget.

Albers and Grinspoon (1997) compared two monitoring and enforcement regimes and identified clear trade-offs. Increasing the budget for enforcement allows both for a larger area to be monitored and for an increased reliance on "police and punish" approaches to enforcement, as for example in the Khao Yai National Park in central Thailand. In contrast to the more inclusive but more poorly-funded approaches in the Xishuangbanna Nature Reserve in south-western China, these approaches are effective in deterring some encroachment, but also induce local people to undertake socially costly avoidance activities. Higher monitoring and enforcement intensity in Khao

Yai National Park did not result in uniformly superior conservation outcomes, however. While the core areas of the national park are better protected from resource extraction and agricultural encroachment, these activities are pushed to the edge of the national park where the risk of detection and punishment is lower. By contrast, the negotiated approach in the Xishuangbanna Nature Reserve results in lower conservation achievements, but greater ability to influence the spatial patterns of activities (Albers and Grinspoon, 1997).

These findings mirror a general conclusion of the literature on conflicts surrounding protected areas; namely that effective policing of protected areas involves considerable costs and an ongoing funding commitment that need to be allowed for in the planning process (Neumann, 2004; Peluso, 1993).

Costs of conflict

There is increasing evidence that conflicts over natural resources can have significant costs that are relevant even at the macroeconomic level (Sachs and Warner, 1997; Bannon and Collier, 2003). This is particularly the case when resources are spatially concentrated, *e.g.* for biodiversity policies which have spatially explicit targets in the form of ecosystems and habitats (Bulte *et al.*, 2005).

The economic literature provides two lenses through which to view the problem of conflict: *i)* rent-seeking models, where the size of the rent to be captured and the cost of acquiring this rent are usually fixed; *ii)* conflict models, where the size of the rent and the opportunity cost of capturing it are endogenously determined (Wick and Bulte, 2006). Both conclude that conflict decreases welfare since it is unproductive, but the scale of welfare losses to be expected differs (Neary, 1999). This work leads to two general messages:

1. Policy-makers need to foresee policy-induced conflicts and factor their expected cost into the evaluation of policies before ranking them for efficiency purposes.
2. Policy-makers need to recognise that conflicts have spill-overs beyond the specific policy under debate. This is because conflicts destroy the social capital on which existing and future institutions rely (Pretty, 2003).

Box 5.3 illustrates the loss of welfare brought about by policy-induced conflicts in the case of Natura 2000 designations in Finland.

In the context of biodiversity policies, conflicts have been analysed theoretically and empirically, with a special emphasis on forest conservation. In a well-known study, Alston *et al.* (1999) examined conflicts and violence in the wake of land reform in Brazil. There, insecurity of property rights is conducive to conflict, since investing resources in a contest over land is more profitable the higher the probability that the land will fall to the contesting

Box 5.3. **Conflicts between private forest owners and biodiversity policy-makers in Finland during the Natura 2000 designation process**

Every EU member state has to designate areas of European biodiversity significance as part of the European conservation network, Natura 2000. In Finland the planning process began in the mid 1990s, and the first lists of areas to be included were published in 1997. The national environmental authorities received almost 15 000 letters of complaint. One example of this protest occurred in Karvia, a small community in south-west Finland. Its natural environment is dominated by bog and marshland, but local landowners had drained some areas in an attempt to develop forest. Conservation in these areas is focused on the remaining bog and marshes, some parts of which are already under protection. The Natura 2000 proposal would have increased the total protected area, affecting the properties of many landowners.

There was a strong outcry in Karvia, with four landowners going on hunger strike in protest against the proposed Natura 2000 Network. This got much public attention and ultimately nearly half the areas were withdrawn from the Natura 2000 proposal. A survey showed that the landowners wanted to take an active part in the planning process from the beginning, rather than only reacting to proposals. They wanted to express their own views on the aim of the planning, the alternatives available and on the impacts (Hiedanpää, 2002).

The main lesson from this case is that participation in the planning process could have prevented the conflict. The administrative solution helped solve the potential distributive problem, but ended with a worse result from a conservation point of view.

party. This points to the importance of creating well-defined and enforceable property rights, the lack of which were found to be key determinants of conflict in other studies (Haro *et al.*, 2005; Fearnside, 2003).

The important contribution of unambiguous and enforceable property rights in preventing conflicts was re-emphasised by Burton (2004), who modelled formal and *de facto* property rights. Burton showed that conflicts over policies that restrict access to previous users will take on the character of a "war of attrition" in which the holder of formal property rights can lose to local residents willing to invest in *de facto* property rights (*e.g.* by blocking access). The reason is that by imposing a sufficiently high negative distributive impact, policy-makers make it worthwhile for those affected to invest considerable resources in a conflict, especially where there are no outside options. The formal nature of property rights is therefore less important for the probability of conflict than the credibility of their enforcement.

The immediate policy relevance of these results is evident in recent studies of protected areas in Africa, in particular national parks. Schmidt-Soltau (2003) and Cernea and Schmidt-Soltau (2006) examined the conservation success of parks whose creation had involved resettling local residents. Conservation was generally poor because the local people invested more in defending their *de facto* property rights than government had invested in the credible defence of the formal property right it had accorded to itself in the post-implementation phase. While this implies that for the local population the welfare losses are less than under a fully enforced policy, a count of overall welfare losses now also needs to include the shortfall in the conservation targets and the cost of conflict for both the rural population and the government.

Engel *et al.* (2006) take the argument based on *de facto* property rights further. They studied a forest conservation situation involving three parties (indigenous community, logging companies, outside NGO) and showed that the variability in conservation outcomes between communities can be traced back to the weak property rights regimes in forests. In conflicts between logging companies and indigenous communities, third-party policy interventions intended to strengthen the conservation of forests can improve conservation, but can also have the paradoxical result of inducing more logging.

5.3.6. *Implications*

Separating policies aimed at correcting externalities (such as biodiversity policies) and policies aimed at redistributive objectives is a key doctrine of welfare economics. Its aim is to maximise the social gains available for redistribution after the policy has been carried out. However, for biodiversity policies there are a number of fundamental and practical reasons why such separation is not possible, and why pursuing biodiversity policies without considering their distributive consequences may involve serious efficiency losses. Some of these efficiency losses will be very palpable, as in the case of conflicts arising around conservation policies (see the German case in Box 5.2). Others will be large, but only evident to future generations. Some may be difficult to foresee at the time of planning, since groups, individuals and institutions will change their behaviour over time. Whether these efficiency losses will be of sufficient scale to sacrifice separation and warrant an explicit distributive orientation of biodiversity policies cannot be answered at a general level. But a general conclusion is that policy-makers need to be aware of these possible consequences.

 PEOPLE AND BIODIVERSITY POLICIES – ISBN 978-92-64-03431-0 – © OECD 2008

The first implication of non-separability is that deciding between policies by simply ranking criteria based on a score of the net present social value is no longer sufficient. This is because those policies generating the highest score are not necessarily those with the most desirable distributive effects. In fact, there is evidence of a trade-off between highly effective conservation policies and equity impacts (Barrett *et al.*, 2005).

The second implication is that an approach is needed for incorporating distributive outcomes into biodiversity policies. This would require policy-makers to:

- Gather information about expected impacts.
- Define criteria for meaningful measures of distributive impacts so that these impacts can be represented and communicated internally and externally (Part I offers a wide range of criteria that are accepted measures of distributive impacts and points out their advantages and drawbacks).
- Define criteria for weighing different distributive issues.

The third implication is that rules and procedures are needed for choosing between different policies with different levels of efficiency and different distributive outcomes. As Part I makes clear, policy objectives and instrument choice are key determinants of distributive effects. However, they are also determinants of the net social gains from these policies.

For policies to be defensible on broad welfare grounds, net social gains need to remain at the core of the policy choice process. Efficiency, in other words, needs to be preserved to maintain legitimacy of biodiversity policies as primarily addressing externalities in habitat and ecosystem management. However, some rules about how to trade off efficiency gains and distributive effects have to guide the process when both dimensions matter. Additionally, in order to be politically acceptable, these rules need to be meaningful, transparent and widely applicable.

5.4. Integrating efficiency and equity into biodiversity policies

5.4.1. *The key issues*

If, for political or ethical reasons, it is decided that separating a policy's efficiency and equity impacts is not feasible, this leads to complexity in policy-making. A policy that needs to consider effectiveness as well as distributive consequences is inherently more complex than if only effectiveness needs to be considered. However, some of the measures for bringing distributive issues into policy decision-making are easily implemented and increasingly part of best-practice in environmental policy-making. Other measures are considerably more involved but should also be part and parcel of good practice in policy-making. They will be discussed further below.

5.4.2. Some solutions: making distributive outcomes matter in biodiversity policies

We offer four approaches for integrating distributive issues into biodiversity policies:

1. Methodological: maintain the linear policy model, but better integrate the welfare impacts on different groups through explicit weightings when determining the potential aggregate welfare improvement that a policy could bring about.
2. Procedural: enrich the policy-making process with numerous feedback loops at different levels of decision-making that involve those individuals who will be directly affected by biodiversity policies. At the lowest level, this starts with simple measures to enhance information flows between policy-makers and individuals and groups. This aims to ensure that those affected will have the greatest possible scope for communicating the expected impacts and influencing decisions as the policy is formulated and the policy instruments chosen. This approach gives rise to the various case studies in Part III of consultative and participatory approaches for addressing distributive issues and resolving distributive conflicts.
3. Institutional: accompany biodiversity policies with explicit changes to the institutional structure under which individuals and groups take decisions that affect the target habitats and ecosystems. This approach gives rise to the various case studies on institutional solutions to distributive issues in Part III.
4. Combine the second and third approaches to bring about institutional changes to allow affected individuals, households and groups to become involved in policy decision-making on an ongoing or even permanent basis. In its most extensive form, this includes measures that devolve to individuals or groups affected part of the management of the policy.

We discuss each of these approaches in detail in the following section.

Methodological solutions: integrating distributional effects into policy design

Part I included a wide range of methods to assess the distributive effects of biodiversity policies. As a reminder, three groups of methods were defined: *a)* methods based on income-equivalent measures (summary measures of equality such as the Lorenz curve, extended versions of CBA, social accounting matrix, distributive weights and Atkinson inequality index); *b)* alternative measures (employment based analysis and child health based analysis); *c)* multidimensional measures (stochastic dominance analysis, multi-criteria analysis and social impact assessment). These methods help the policy-maker

PEOPLE AND BIODIVERSITY POLICIES – ISBN 978-92-64-03431-0 – © OECD 2008

identify the main groups affected by the policy and the important distributive effects in monetary and social terms.

The first and most basic approach to integrating distributive effects into policy-making is to give these measures an explicit role in computing the potential aggregate welfare improvement of policies and in choosing instruments. This means that the policy-making process is now augmented by a consideration of distributive impacts. At the same time, the policy-maker still retains full control over information gathering, policy evaluation and choice, and instrument choice. Trade-offs or possible win-win situations between the main goals of the policy (biodiversity effectiveness, economic efficiency and equity) will thus be revealed to the policy-maker and can become the basis for decision-making. A graphical way of depicting the linear policy model augmented by distributive considerations is shown in Figure 5.2.

Each of the methods discussed in Part I is, in principle, suitable for augmenting the linear policy model. Those methods that allow the policy-maker to rank policies in a way that is sensitive to distributive impacts (such as a CBA with distributional weights) are naturally the easiest in procedural terms. At the same time, policy-makers may feel more comfortable with methods that are less explicit and less mechanical in integrating distributive effects, such as multi-criteria assessment.

The benefit of this approach for the policy-maker is that they can drive the entire policy-making process. Responsibility does not have to be shared with other policy participants and the policy-maker is able to interpret information freely.

Figure 5.2. **Linear policy model adapted to include distributional measures**

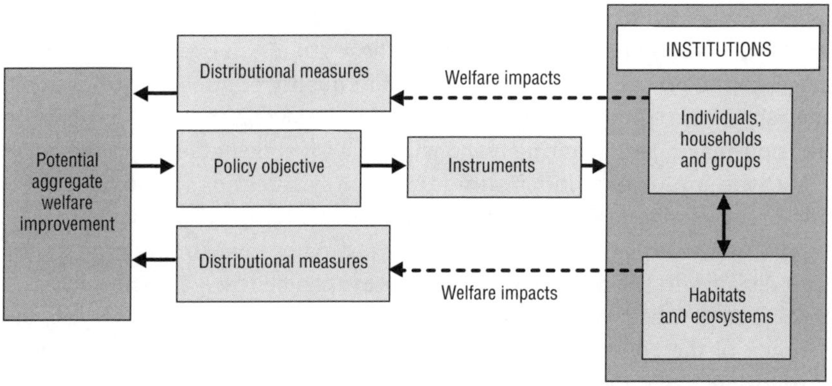

Procedural solutions: making distributive issues explicit through the biodiversity policy choice process

The second solution to integrating distributive concerns into biodiversity policies is procedural. Its main idea is that equity issues are best addressed through a mixture of measures that enhance the policy-making process.

One way to think of procedural solutions is as an escalating scheme of stakeholder group involvement. At the simplest level, procedures should be enhanced in order to address the information gaps identified in this chapter. These gaps partly concern the information flow from stakeholders to policy-makers. This means that when the welfare impacts of different policy objectives and instruments are assessed and policies ranked, the relevant information on which the impacts, and hence the ranking are based, is incomplete. Clearly, resolving information gaps at that time is not only a way of bringing distributive effects to the attention of policy-makers, but also of enhancing the overall efficiency of the policy-making process.

The other gap is in the information flow from policy-makers to stakeholders. A well-established source of conflict over biodiversity policy implementation is when stakeholders are informed too late and with too little detail about the exact nature and timing of the chosen policy. Robust communication strategies allow local stakeholders to consider all the relevant facts in order to decide on the best response. The communication strategies are depicted in Figure 5.3 as information flows that connect biodiversity outcomes and individuals with the policy-making process.

Communication strategies alone do not guarantee that the information and concerns that stakeholders bring into the process will influence the outcome of that process. A higher level of involvement in the decision-making process opens up to stakeholders the bodies that choose the policy objectives and/or the policy instruments and gives these groups an explicit vote at these stages of the process. This brings a different quality of impact information to the table, leveraging the information process for enhancing the quality of decision-making and sharing responsibility with stakeholder representatives about the important policy dimensions. Key questions that have to be addressed at this stage are what type of decision-making involvement is representative, what weight stakeholders should be given in the process, and how disputes in the decision-making process can be resolved. The successful resolution of these conflicts is a key condition for ensuring the integrity and viability of the policy process. Part III provides detailed guidance on how to prevent, manage and settle disputes that arise out of distributive issues in biodiversity policies. The involvement of individuals in formulating policy objectives and choosing instruments is depicted by solid arrows in Figure 5.3, thus forming feedback loops at two points in the policy process.

Figure 5.3. **Linear policy model adapted to include procedural focus**

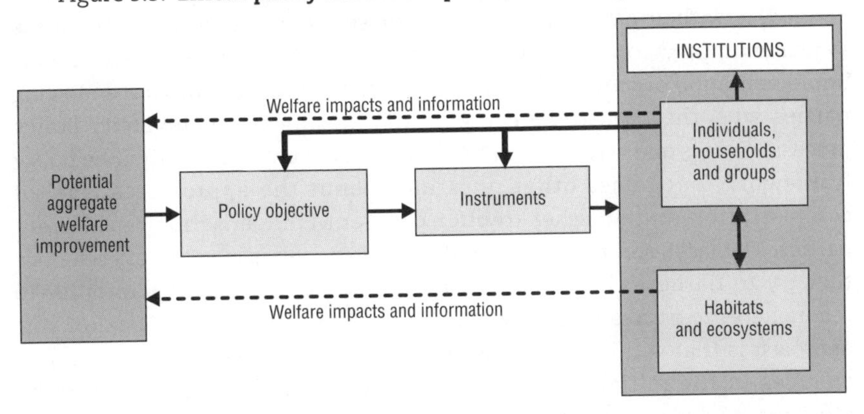

Institutional solutions: making distributive issues explicit through institutional changes

The third approach to addressing distributive issues in biodiversity policies focuses on the role of institutional change as a conditioning factor in policy implementation. This approach assumes that biodiversity policies and their instruments are accompanied by explicit changes to institutional structure. The causal pathway relies on the importance of these institutions for individual and group decision-making on activities affecting local habitats and ecosystems (Figure 5.4).

Figure 5.4. **Linear policy model adapted to include institutional focus**

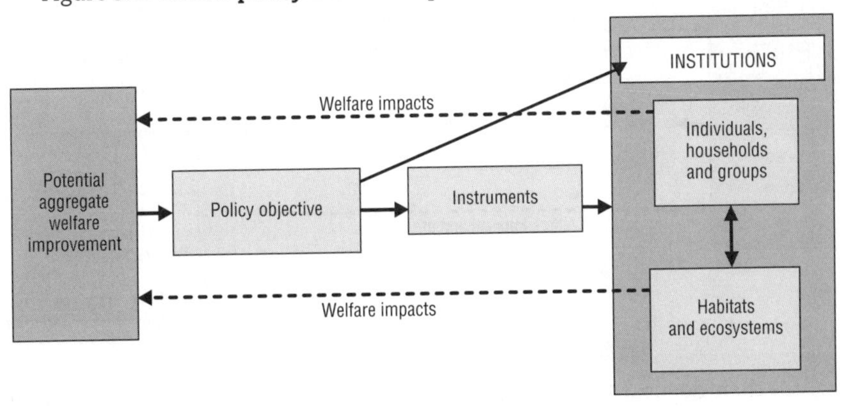

In Part III therefore, we use a rich set of case studies to show how policy-makers can succeed in "getting the institutions right" (Barrett *et al.*, 2005). We explore how to identify those groups that should benefit from institutional changes alongside biodiversity policies; how these new rights should be defined, how their new "owners" can exercise them, and what experience is available for predicting their distributive effects. We also look at very specific implementation of property rights, such as compensation titles (*i.e.* endowing parties with the right to be compensated during the biodiversity policy process). This is one way of accommodating individual losers within the policy framework, but raises other questions about the appropriate level of compensation and whether to offer different compensation to different parties. The last section draws from the rich experience in voluntary schemes that try to harness the powers of market-based mechanisms in order to reconcile efficiency and equity objectives. The intellectual appeal of this approach is that within limits, it is possible to address equity issues through changes in the initial endowment positions without compromising the efficiency of the market later.

A combined approach: procedural and institutional change

An augmented biodiversity policy-making process that integrates distributional objectives into objective formulation and instrument choice (Figure 5.5) will change property rights, use rights and procedural rules in order to alleviate distributive concerns in the application of the biodiversity instrument.

Figure 5.5. **Linear policy model with procedural and institutional focus**

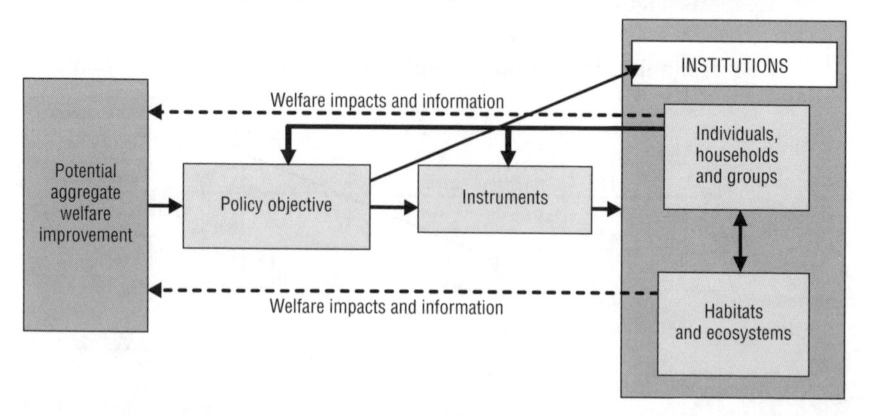

In their deepest form, the procedural and institutional approaches grant affected groups and individuals the right to ongoing involvement in the decision-making process. A common form of this approach is to devolve management rights down to the level of stakeholders or their representatives. This gives those most affected by the policies a considerable role in managing the day-to-day implementation of the policies. It requires as a precondition that stakeholders buy-in to the policy objectives and the instrument choice. On the other hand, it also means that the policy-maker relinquishes control over important parts of policy implementation. The requirements on both stakeholders and policy-makers are therefore considerably higher in this model than in the two preceding procedural solutions. Part III offers various empirical examples of so-called "co-management" in which this form of involvement has succeeded.

5.5. Summary and conclusions

We started this part by asking a fundamental question: Given that biodiversity policies have distributive effects that are real, non-marginal and can be measured, should these effects matter for policy choice? A central "doctrine" of policy choice is that policies targeting externalities (such as biodiversity policies) should best be left to get on with internalising these externalities as much as possible, thus creating the greatest possible social gains. Distributive effects of the many policies simultaneously pursued by governments will in all likelihood even each other out. Should non-marginal effects persist in the end, then efficiency is still best maintained by lump sum transfers from winners to losers after the policy is implemented, rather than burdening policies with distributional objectives. Separation of equity and efficiency should be the norm.

As a response to this argument, in this part we have reviewed the fundamental and practical challenges to separating these aspects in biodiversity policies. We have shown that the theoretical justifications for separability in the case of public goods are not strong and that the presence of information imperfections and transaction costs in transfers severely compromises separability. Similarly, the inability of institutions to bring about the lump-sum transfers to compensate for non-marginal changes hinders separability. Reasons for this inability are the spatial and intertemporal limitations of jurisdictions; political economy and differential access to power; the contestability of property rights through conflict; and finally the institutional endogeneity of efficiency and equity in situations where core elements of the habitat or ecosystem are managed as common property resources.

The main conclusions of Part II are that distributive issues matter for biodiversity policy-making for a number of quite fundamental reasons. In fact, paying attention to distributive outcomes will often enhance efficiency. Policies built on excessively narrow definitions of efficiency can often lead to wasteful conflict and can be ultimately self-defeating. However, distributive effects have to be sufficient to forego the benefits of leaving biodiversity policies unburdened with accomplishing potentially efficiency-reducing distributional objectives.

When distributive effects are significant, biodiversity policies should not lose sight of efficiency objectives while attempting to mitigate their distributive outcomes. There are four main options for integrating distributional concerns into biodiversity policies: explicit distributional weighting of policies; improving the policy-making process by enhancing communication with and participation of stakeholders; implementing institutional changes alongside biodiversity policies; and combining the second and third approaches to bring about institutional changes to allow affected individuals, households and groups to become involved in policy decision-making on an ongoing or even permanent basis. The last three approaches are the subject of Part III.

PART III

Bringing Distributive Issues into Biodiversity Policies in Practice

ISBN 978-92-64-03431-0
People and Biodiversity Policies
Impacts, Issues and Strategies for Policy Action

Chapter 6

Procedural Approaches: Communication, Participation and Conflict Resolution

6.1. Introduction

At the end of Part II four approaches were discussed to integrate distributive issues into the national biodiversity policy-making process:

- Methodological approaches: integrating the measured distributional effects into policy design.
- Procedural approaches: communication, participation and conflict resolution strategies.
- Institutional approaches: compensation schemes and voluntary approaches.
- A combination of procedural and institutional approaches: ongoing involvement of local communities and other stakeholders in management decisions.

Having discussed the first point in Part I, in Part III we describe how to put the last three points – procedural, institutional approaches and their combination – into practice, mainly for national policies. International policies are noted briefly for the channels through which they are relevant, and for the magnitudes of the transfers occurring. In each chapter the main issues are introduced then descriptions and comparisons of different solutions are given, followed by some illustrating cases. The intention is to encourage policy-makers to take steps towards using participatory methods for co-operating with important stakeholders affected by policy. By doing so, considerable distributive effects can be prevented, mitigated or made acceptable at an early stage of the policy process.

This chapter describes how to address distributive issues through communication with, and involvement of, stakeholders during the decision-making process. With the exception of conflict management, these approaches should precede the implementation of the policy. They help improve the policy formulation process, and help choose policy instruments so that distributive issues can be integrated at an early stage. If successful, they prevent conflicts from arising. If conflicts do arise during policy design or implementation, procedures can be put in place to help resolve them. The last part of the chapter covers procedures used in dealing with and resolving conflicts.

6.2. The value and implications of communication and participation

Important characteristics of effective participation include:

- Providing facts and technical information in an understandable form (using non-technical language, illustrative charts and examples).
- Using appropriate communication channels: newsletters, articles, TV-radio news, internet, forums.
- Providing opportunity for feedback and discussions (through letters, email, internet, phone, forums, roundtables).
- Providing information relevant to specific stakeholder groups (emphasising what positive and negative impacts may be felt by the group and how they are balanced).

Effective consultation allows various groups to express their views so potential conflicts can be addressed and acceptable solutions developed. The process of effective stakeholder dialogue has some specific characteristics (Declerck *et al.*, 2003):

- Seeks to find common ground.
- Aims to provide a result everybody can live with.
- Is structured so that process is as important as outcome: builds ownership of outcomes.
- Has no predetermined outcome.
- Engages those who will be affected by the outcome at the beginning of the process.
- Is collaborative, working with, rather than for, people.
- Engenders ownership of solutions and a commitment to their successful implementation.
- Gives all stakeholders a voice, including local people.

Other principles that have been shown to give positive results in deliberative processes include: involving people in a timely manner and giving them enough time to express their views; good facilitation; incorporating the results into decisions; and flexibility in using different procedures (Carson and Gelber, 2001).

Participatory methods engender strengths that policy-making should attempt to harness; but of course they also create obstacles that must be overcome. A review of those strengths and weaknesses is given in Table 6.1.

While undoubtedly useful and productive, the benefits of the various procedural approaches must be weighed against their cost in both time and resources. Given that consultation requires considerable time and effort by

Table 6.1. **Strengths and challenges of participatory methods**

Strengths	Challenges
Understanding of policy issues can be increased.	Good preparation and skilled moderator/facilitator are required.
Acceptability of the policy can be increased.	Representativeness of public is sometimes hard to achieve.
Conflicts can be prevented.	Public officials need to re-orient their perspective away from authoritarian attitudes.
New creative ideas can emerge during the process.	Costs vary, for example experts need to be paid.
Implementation of the policy measure will be smoother.	Good briefing and introductory material is needed.
Expert knowledge can be combined with public and stakeholder opinion to find the best solutions.	Time lags can be considerable (months of preparation required even for simple processes).

government staff, *e.g.* travel expenses, there is a need for cost/benefit considerations even in planning a participatory process.

Addressing the interests and needs of communities and stakeholder groups, and consulting with these groups, requires at least minimal capacity to respond and to adapt policies. Timelines for policy implementation must thus explicitly accommodate such interaction. By implication, policies that are more disruptive and more likely to have significant redistributive impacts will have to have longer timelines – the need for providing information, getting feedback and making adjustments will be greatest. In other words, one should expect to see a direct correlation between the magnitude of a policy and the amount of time spent in the preparatory stages.

If time and cost issues are not accounted for, this might limit public communication and consultation to information campaigns through the media, calls for public comments (to be submitted to the ministries or authorities in a written form), public hearings or consultation. These forms of public involvement are usually reactive in nature (Konisky and Beierle, 2001), and the problems they uncover may be difficult to address within a fixed agenda. This can leave the general public and stakeholders dissatisfied with the outcome and the process.

6.3. General methods for public involvement

At a practical level, there are many participatory procedures available to policy-makers (Box 6.1, and see OECD, 2002). Below we explore some of these procedures and assess them for the circumstances in which they work best and how they can address distributive issues. We distinguish between those that engage the general public *versus* those that focus on specific stakeholders, although there is considerable overlap between the two.

Different procedures suit particular circumstances so attention needs to be given to the relative strengths of each. They can differ on issues such as the

PEOPLE AND BIODIVERSITY POLICIES – ISBN 978-92-64-03431-0 – © OECD 2008

Box 6.1. **Methods of public involvement**

- **Search conferences:** conducted at the beginning of a planning process. A small and knowledgeable group establishes a long-term vision and develops long, medium and short-term actions to reach that goal. The group is not representative; individuals are selected on the basis of their knowledge and constructive collaborative ability. The group meets only once for one or two days. Search conferences cannot substitute for broader public consultation but can serve as a preparatory work for it. The creative outcome of the search conference can feed into broader consultation on a complex policy issue. There are many situations where it could be used in biodiversity policies: *e.g.* land-use planning where biodiversity aspects need to be integrated, development of agri-environmental schemes for a region or for the whole country, or the introduction of economic incentives. Distributive issues can be included, but the group needs to be specifically asked to do this.
- **Deliberative polls:** a large sample (perhaps up to 500 people) is invited to a special location for several days to discuss an important and sometimes controversial policy issue. The group is supposed to represent the community and be large enough to have a significant result. The whole group is subdivided into smaller groups to determine the issues to be discussed. Moderators/facilitators are involved to channel the discussions and help in the procedure. Participants vote at the end, and shifts in opinion during the process are examined. Deliberative polls could be used in biodiversity policy planning, especially for nation-wide, complex and contradictory policy issues, *e.g.* discussing user (hunting, logging or fishing) rights in protected areas, regulating the collection and trade of protected plant species, or problems with genetically modified organisms. Distributive issues can be made implicit in the discussion.
- **Citizens' jury or community panel:** a small number of individuals is usually randomly selected from the general public to form a jury and asked to deliberate on a policy issue (usually coming from an agency). Participants meet for two to four days and they are given briefing material in advance. They are presented with different options by experts on different aspects of the issue (*e.g.* financial, biological, legal, social or ethical aspects). Moderators or facilitators conduct the discussion and help resolve conflicts. At the end of the session a report is prepared listing the jury's recommendations. Using this method in biodiversity policy can include *e.g.* discussions on the development options for a nature reserve or protected area, zoning within and around a protected area, or preparation or revision of a management plan for a protected area. Distributive issues are usually part of the process, especially when discussing different options for action.

Box 6.1. **Methods of public involvement** *(cont.)*

- **Consensus conference:** a panel of a small number (around a dozen) people who are set a specific question, usually on a broad-ranging scientific or technological issue. The consensus conference is very similar to the citizens' jury, but it takes place over a longer period of time and generally involves preparatory weekends. The conference usually has a professional moderator, whose role is to facilitate the dialogues and resolve conflicts. Experts representing different opinions are interviewed (*e.g.* relevant stakeholders, interest groups, NGOs, technical experts). A final report is written at the end and submitted to the agency in charge of the policy. In many countries it has good media coverage. The biodiversity related themes can be similar to those of the citizens' jury, *e.g.* policies on genetically modified organisms, development options for a certain natural area, establishment of a protected area, discussion on the access and users rights connected to a natural area, action plans against invasive species. Distributive issues can be included in the process and often are naturally part of it.
- **Charette:** an intensive, consultative planning process over about five days involving a rapid and dynamic interchange of ideas between planning practitioners, stakeholders and the general public. A team of planning practitioners prepares and publicises discussion material on the issue. The first day the team meets to draw up some preliminary issues for discussion then a public meeting is held collectively and in smaller facilitated groups. The next day, the team meets the stakeholder groups to discuss the issue. The following day, the team puts together a list of options by combining their understanding of community concerns and stakeholders' needs. The options are prepared in a format that is open to public inspection. Follow up meetings might be held. This method can be used in biodiversity policy, *e.g.* for discussing the development options for a nature reserve or protected area, land use planning with the inclusion of biodiversity aspects, zoning or preparation of a management plan for a certain protected area or nature park. Distributive issues are usually naturally included in the discussion (many interests are confronted during the sessions).
- **Residents' feedback panel:** established from a pool of potential respondents in a given area, who are called upon for surveys, interviews or consultation for an issue where public opinion is needed. The panel operates for two to four years. It can also be used in connection with biodiversity policy, *e.g.* long-term development of a natural area, development or revision of a management plan for a protected area, involvement of locals and indigenous people in the decision-making for a protected area, development of a set of economic instruments for biodiversity policy. Distributive issues can be included in the set of questions under discussion.

Source: Carson and Gelber (2001) with applications to biodiversity policies and related distributive issues.

PEOPLE AND BIODIVERSITY POLICIES – ISBN 978-92-64-03431-0 – © OECD 2008

specifics of the policy at stake, the budget available, the stage of the decision-making process and their potential to include distributional effects. Table 6.2 can help guide the choice of the best method.

Table 6.2. **Comparison of participatory methods**

	Search conference	Deliberative polls	Citizens' jury	Consensus conference	Charette	Residents' feedback panel
Do participants determine key questions?	Yes	Not usually	No	Yes	Yes	Not usually
Random selection?	No	Yes	Yes	Yes	Yes	Not usually
Number of participants	20-50	Several hundred	12-25	12-25	Up to several hundred at public meeting, up to 20 in stakeholder meeting	From 50 to several thousand
Do participants meet?	Yes	Yes	Yes	Yes	Yes	Not necessarily
Time involved in face to face meetings	1-2 days	1 to 3 days on site	2-4 days	2-4 days on site plus 2 preparatory weekends	2-5 days	Can be undertaken without face to face meetings
Time to findings	A few weeks to a few months	6 months	2-6 months	12 months	Several weeks	RFP exists for 2-4 years and called upon many times
Type of outcome	Long-term vision, broken down into short-term action plans	Votes recorded before and after deliberation	Written report of findings	Written report of findings	Planning proposals, with sketches and maps if appropriate	Usually quantitative survey data
Are findings published in a report?	No	Yes, by commission-ing authority	Yes, by commissioning authority	Yes, by participants	No	Yes, by commissioning authority
Are experts brought in as witnesses?	No	Yes	Yes	Yes	Yes	No
Key issues	At early stage, to set parameters for plan making in a region	More informed opinion poll	Complex issues requiring lengthy deliberation	When process can be opened up for public input, issue is complex	Intensive, fast planning decisions with community involvement on a specific issue	Track changes over a long period of time, use as database for other consultative methods
Opportunity to address distributive issues in the discussion	Yes, but need to be included in the agenda	Yes, but need to be included in the agenda	Yes, naturally included	Yes, naturally included	Yes, naturally included	Yes, but need to be included in the questionnaires

Source: modified from Carson and Gelber (2001).

6.3.1. *Stakeholder involvement methods*

There are also methods more suited for discussions with specific stakeholders, rather than the general public. These include joint fact-finding, focus groups, discussion forums, roundtables, scenario workshops and negotiated rule-making (Box 6.2). In these forums, certain interests can be expressed and potential conflicts can be resolved before a policy is introduced. Definition of the methodology, however, is not as well developed and boundaries between the methods are not so clear. There are also differences between the procedures in how and in what depth they are suited to discuss distributional issues.

Table 6.3 summarises the key elements of the methods listed in Box 6.2.

6.3.2. *Examples of participatory methods*

Some examples of using these procedures for biodiversity and nature conservation are summarised in Table 6.4, and a few of them are discussed below. Some of them have been used for research purposes; others were used in actual decision-making processes.

Citizen's jury for wetland management in the UK (summarised from Aldred and Jacobs, 2000)

Ely's citizens' jury was organised in 1997 in Norfolk in the UK to discuss four wetland management scenarios. The jury consisted of 16 members of the local public. The four options were the following:

- Option 1. A nature reserve (4 800 hectares incorporating rare wildfowl and mammals).
- Option 2. A fen centre (multi-use recreation and tourism centre).
- Option 3. Incremental development (wetland creation through small-scale, farmer-led initiatives).
- Option 4. No deliberate option.

Each of the first three options was a genuine proposal seeking public funding. The jury was given short presentations by experts on different aspects of the question. The jury was regularly asked to split into small groups, whose composition changed each time. Each group appointed a spokesperson to report back their discussion to a plenary session of the whole jury. Conclusions: no single option was favoured – Options 1 and 3 were both supported. The nature reserve (Option 1) was strongly supported on the grounds that rare species should be protected. However, there was disagreement about the size of the reserve and discussion about alternative sites. The jury suggested incorporating educational and recreational activities into Option 1. Option 3 (incremental development) was also supported on the

Box 6.2. **Specific stakeholder involvement methods**

- **Focus groups:** usually the representatives of only one stakeholder group (or stakeholder groups with similar interests) are invited to express their perceptions of and interests in the proposed policy. Focus groups are often conducted as a preparation for citizens' jury or roundtables to collect prior information on the case and on the potential conflicts. They can be used when certain groups are affected by a biodiversity policy measure: before the preparation of a management plan, zoning, restriction of certain user rights, incentives for nature friendly land management, or modification of hunting or fishing rights. The main interests and concerns of the group can come up during the focus group discussions, and they almost always have distributive elements as well.
- **Discussion forums/roundtables/scenario workshops:** fora in which representatives of different stakeholder groups can discuss possible policy options and propose solutions to decision makers. Good facilitation is needed to channel the discussion, to help resolve potential conflicts and to allow different views and interests to come to the surface. They are widely used and have great potential in biodiversity policy planning, *e.g.* in the planning phase of a new regulation, development of a national park, or before the introduction of agri-environmental measures. They are useful for addressing distributive issues that are important to specific stakeholder groups.
- **Negotiated rule making:** In this process, an administrative agency convenes representatives of the regulated economic sector, public interest groups and other stakeholders to seek agreement on either the elements of, or specific language for, a proposed regulation, prior to initiating notice and comment (*www.resolv.org/tools/concepts.html*). It can be used in biodiversity policy making as well, *e.g.* before a biodiversity regulation is introduced or a management plan for a protected area is finalised. This is the last chance for the stakeholder group to present their interests and have them included in the proposed regulation. The main discussion on distributive issues needs to have taken place beforehand.
- **Joint fact finding:** helps deal with the technical complexity of issues and with scientific uncertainty, where this creates obstacles to an agreement. Parties discuss what factual questions they believe to be relevant to the decision, exchange information, identify where they agree and where they disagree, and negotiate an approach to seeking additional information, either to fill gaps or to resolve areas of disagreement (*www.resolv.org/tools/concepts.html*). It can be used in almost any biodiversity policy-making process, *e.g.* preparing new nature conservation regulations, introducing new biodiversity policy instruments, drawing up action plans for reducing

Box 6.2. **Specific stakeholder involvement methods** *(cont.)*

new biodiversity policy instruments, drawing up action plans for reducing invasive alien species. Distributive issues can come up when data on the extent of the distributive effect on different economic or social groups are needed, or when certain arrangements need to be finalised for settling distributional problems.

Source: Carson and Gelber (2001); *www.resolv.org*; Andersen and Jaeger (1999) with applications to biodiversity policies and related distributive issues.

Table 6.3. **Summary of stakeholder involvement methods**

	Joint fact finding	Focus groups	Discussion forums, roundtables, scenario workshops	Negotiated rule-making
In which part of the planning process are they used?	Usually at the beginning of the policy process	At the beginning or in the middle of the policy process	In the middle of the planning process, when the goals are set	Before the finalisation of the policy
Time involved in face to face meetings	Depends on the complexity of the issues: half a day or more rounds	Few hours to a day for each group	One or more days	Half a day to a day
Key issues	To discuss scientific facts, uncertainties and identify ways to get more information	To determine main problems, interests of the stakeholders, some proposed action	To discuss the goals, problems, solutions with the stakeholders	Specific language of the proposed regulation or policy
Type of outcome	Identification of areas where more information is needed and identification of potential areas of conflict	Opinions and views of the key stakeholders	Proposal for decision-makers	Final regulation or policy
Facilitation	Optional	Yes	Yes	Optional
Opportunity to address distributive issues	Yes	Yes (but concerning only one stakeholder group at a time)	Yes	To a limited extent

grounds that it is important to provide habitats for wildlife all over the fens, not only in separate sites, and that significant effort could be made by landowners. This approach would build on and extend existing initiatives and good practice, while joining the initiative would remain voluntary. In addition to the proposed four options, a new option was suggested by the jury: for a local wholesale centre to distribute fruit and vegetables produced in the fen (titled: Fens' Covent Garden). It thought that the job creation and economic

Table 6.4. **Synopsis of cases**

Type	Example	Distributive issues
Citizens' jury	Wetland management, UK	Different options with different distributive effects: nature reserve/development of tourism/small-scale farming
	National park management, Australia	Financing the management of the park through a levy (progressive or not?)
Focus groups	River dialogue, Sweden, Netherlands, Estonia	Some conflicting interests and potential measures (*e.g.* compensation for farmers)
	Wetland valuation, Greece	Conflicting interests in development: fishing/tourism development/nature conservation/agriculture
Roundtable, national workshop	The Boreal Forest Program, Canada	Different interests of the extractive industry, non-governmental organisations and Aboriginal organisations
	Designation of critical habitat, USA	Economic impacts of designation, incentives

development potential of the centre would be significant and it would help products coming from the fens to stay in the region.

This case shows that nature conservation measures can be accepted by the public if they are introduced in a participatory manner. Distributive issues (allowing recreational activities, using small-scale farming or job creation) can also be discussed during the sessions and selected people may be able to find good solutions for the whole community.

Citizens' jury on national park management in New South Wales, Australia (James and Blamey, 2000)

A citizens' jury was organised by the Australian National University in Canberra, Australia in 1999 to discuss limited-budget management activities for national parks. The organisers conducted focus groups as preparation to help construct different scenarios for the jury. The members of the jury were chosen from the population of New South Wales, and were representative of the population in terms of gender, age, place of residence, rating of environment in relation to other social issues, occupation, income, income source and education. Witnesses were selected both for their technical expertise and for their presentation skills and were expert in fire management, weed control, tourism, recreation, feral animal control, management of historic sites and research. The jury had to decide among three options for allocating the yearly budget of the National Parks and Wildlife Service across the five major park management programmes (Table 6.5).

Table 6.5. **Management options for national parks in NSW**

Outcomes of national park management	Option 1 (current situation)	Option 2	Option 3
Number of national parks with good fire management	100	40	160
Area of feral animal control each year (hectares)	50 000 ha	100 000 ha	30 000 ha
Area of weed control each year (hectares)	3 000 ha	1 000 ha	10 000 ha
Proportion of facilities that are well maintained	35%	45%	25%
Number of well protected historic sites	7 000	6 000	7 500

The jurors selected Option 1 after broad discussion, but they recommended that more funding should be allocated for improving the management of the national parks. The second task for the jurors was to decide how to finance the parks' management. A levy on income tax (paid each year) was proposed by the project team to increase the amount of money available. The jury had a constructive debate but was unable to reach consensus on this issue. Some of the results of the internal discussions were as follows: nine to four voted in favour of the "better park with levy" option). After discussing how to calculate the levy, the pro-levy jurors favoured a progressive levy, calculated as a percentage of gross income. After discussion, two different percentage figures were proposed: 0.1% and 0.25%. Most pro-levy jurors voted for 0.1%. If this was accepted, an additional AUD 109.7 million would have been collected annually for national park management (James, 1999; James and Blamey, 2000).

Financing a national park from citizens' taxes is a distributive issue. The case shows that a citizens' jury can suggest an economic incentive, in this case a progressive levy, which might be more acceptable to other citizens, given that the jury is a sample of potentially affected citizens.

River Dialogue: focus groups in three European countries under the EU Water Framework Directive (Googch et al., 2003; River Dialogue Newsletter 1, 2003)

River Dialogue was an EU research project in 2003-2004. Its objective was to identify the best approaches to increasing public involvement in the implementation of the EU Water Framework Directive and river management plans. The main methods involved focus groups and citizens' juries (Table 6.6) in three European river basins: the Motala Ström in Sweden, Ijsselmeer in the Netherlands and the Emajõgi River in Estonia.

Some characteristics of the focus groups are summarised in Table 6.6. In the second phase of the project citizens' juries were organised in each country.

Table 6.6. **Characteristics of focus groups in River Dialogue**

Characteristics	Sweden	Netherlands	Estonia
Number of focus groups	8	9	9
Participants	Ordinary citizens, sailors/ water recreation interests, farmers, fishermen, local authorities, nature conservation groups	Farmers, fishermen, water recreation, nature conservation groups, homeowners, public officials, citizen groups	Environmentalists, schoolchildren, owners of holiday homes, fishermen, farmers, officials from local authorities, water recreation groups, NGOs, people from the canoeing centre

The main conclusions of the focus groups were the following:

- **Sweden:** there was a difference between the groups which felt more directly affected by water-related issues (*e.g.* fishermen, farmers, local government officials) and those which did not feel particularly affected (*e.g.* ordinary citizens and homeowners). Several participants emphasised that water issues are not of immediate concern to many citizens, as Sweden has significant water quantity and relatively good water quality. Increasing awareness of water among the public was highlighted as important. The lack of more established dialogue between the involved parties, for exchanging views and learning from each other, was also mentioned. The EU role in water management was not seen as negative; however, uncertainty over the implementation of the Water Framework Directive was viewed as negative by some participants.
- **Netherlands:** water quality was of interest to all focus groups. Almost all focus groups (except the fishermen) perceived a notable improvement of water quality in recent decades, although litter and discharge of untreated wastewater were seen as problems. The Ijsselmeer was considered an important natural area by the participants. The members of most focus groups noted that economic developments and nature could go hand in hand, but some groups, *e.g.* nature conservationists and citizens of Friesland, had serious doubts about it. There was widespread support for the European role in water management, but some concerns were raised about its implementation. The issue of regulation was of interest to more groups and was closely related to the organisation of water management and nature. The lack of public involvement in water related policy-making was emphasised by the groups.
- **Estonia:** poorly regulated water transportation was seen as a threat to the ecosystem of the River Emajõgi and to fishermen and swimmers, although water quality was said to have improved. Poorly developed infrastructure, the lack of rubbish bins, camping and parking zones were considered to be

problems. It was emphasised that farmers, who take care of the water meadows (the natural water purification systems and fish spawning areas), should be supported financially. Nature protection institutions restrict traditional and profitable human activities such as agriculture and recreational enterprises on the riverside. Intensive fishing was seen as more of a social problem. Focus groups also showed that the Estonian media does not pay much attention to water issues.

The case shows that river management has a different focus in different regions of Europe. Nature conservation and other uses (recreation, fishing, agriculture, drinking) might be in conflict at some places but participatory methods can reveal the problems and participants can suggest possible solutions. All the solutions have distributive elements, while there are always limits to some uses. The control and the restrictions are likely to be accepted if consensus is reached during the discussions.

Focus groups for wetland valuation in Greece (Kontogianni et al., *2001)*

Focus groups and questionnaire surveys of individuals were used in the Kalloni Bay, Greece, in 1998. Kalloni wetland is one of the most important wetlands in Greece, being one of the country's Natura 2000 sites. It functions as a wintering breeding and migration station for birds. It is also one of the most important fishing grounds in Greece, especially for oysters, and a promising site for the development of aquaculture. Besides its ecological value, the Kalloni wetland is also a tourist attraction with a prominent bird-watching tradition. The wetland is currently under pressure from increased population and the extension of agricultural activities.

Four focus group interviews were held, involving local fishermen, building constructors, hotel owners and elected representatives of the affected villages. Farmers were not interviewed as they were identified as a non-cohesive set of individuals.

- Local fishermen emphasised the value of the bay and the richness of the sea, as well as the importance of preserving it. They considered the problems of overfishing, the pollution coming from agricultural practices and possible negative effects of aquaculture.
- The hotel owners had a lively discussion about the potential for tourism development and the problems of waste disposal in the wetland. In some respects they had a negative perception of the wetland – they thought that it was an unsuitable area for development and that higher water levels could threaten houses. They also felt it was not their responsibility to manage the habitat. Concern was expressed about mass tourism; they favoured the development of small local tourism. There was also a

discussion about the possibility of building a new airport, but they could not agree on the consequences.

- Local elected representatives saw the wetland as an important local resource, and they accepted responsibility for preserving it. They acknowledged the problems of pollution and waste management and uncertain property rights in some parts of the wetland. They all favoured the idea of a new airport.
- Building constructors were mostly concerned about the waste problems and thus pollution of the bay. They extract sand from the bay. Although they realised that it was destructive, they did not see its preservation as their responsibility as it was a legally permitted activity. They also acknowledged the problems arising from the lowering of the water level due to more extensive use of water but they did not take responsibility for this either. They were in favour of future developments, and though they were concerned about tourism and agriculture, they also saw trade-offs between these two activities.

The focus group method – as shown in this case – can reveal the differences in interests, attitudes, and plans of different stakeholders in developing a biodiversity rich area. It can be used as a basis for the analysis of distributive effects, *e.g.* for social impact assessment with stakeholder analysis.

Conserving Canada's natural capital: The Boreal Forest Program – National Roundtable (National Roundtable on the Environment and the Economy, 2005)

The National Roundtable on the Environment and the Economy of Canada examined ways to balance conservation with economic activity in Canada's boreal forest. The programme was guided by a task force consisting of representatives from extractive resource industry sectors, non-governmental organisations, academic organisations and national Aboriginal organisations.

As a result of the programme a State of Debate report, including a set of case studies, was produced. The State of Debate report outlines the current state of Canada's boreal forest, describes best practices and assesses the potential use of regulatory and fiscal policy in further conversation and integrating it with economic activity in the boreal forest. At the end of the programme seven recommendations were made:

- Convene a national leaders' conference on the future of Canada's boreal forests.
- Establish a boreal Network of Centres of Excellence.

- Improve the capacity for climate change adaptation of boreal forests.
- Expand the use of fiscal incentives to promote conservation by resource industries in the boreal forests.
- Strengthen integrated landscape planning and management through innovative approaches.
- Strengthen institutional arrangements of Aboriginal people.
- Support capacity-building of Aboriginal communities.

Three case study regions have been identified: the Muskwa-Kechika Management Area in north-eastern British Columbia, the AlPac Forest Management Area in north-eastern Alberta and the Abitibi region on the Québec-Ontario border. They were identified using the following criteria: pressure of multiple use and conflict; presence of multiple jurisdictions; presence of innovative approaches; incorporation of aspen parklands, taiga and boreal forest; potential for generating forward momentum; and balanced geographic representation. The case studies have been completed and discussed in regional workshops.

This example illustrates how a government can deal with a complex policy issue such as the use of boreal forest in a large country like Canada. A national roundtable is a good forum to discuss the different uses of the forest by stakeholders in different parts of the country and to reveal potential conflicts and potential co-operative actions between the main users. It can also identify the main distributive issues. It is also a useful tool for adjusting existing policy to strengthen conservation and assess new policy instruments (*e.g.* fiscal combined with regulation).

Designation of Critical Habitat – National Workshop Project in the US (Moore et al., *2000)*

In 2000, the US Fish and Wildlife Service held two national workshops to help form new policy and procedures for the designation of critical habitat for species listed in the Endangered Species Act. Participating in the two workshops in Reston, Virginia and Tempe, Arizona, were 28 and 35 invitees respectively, representing different interest groups, regulatory entities and federal agencies with a stake in the direction chosen by the US Fish and Wildlife Service. Observers also attended. The goal of the workshops was to create a forum where specific issues relating to designation could be discussed openly and honestly, and where increased understanding and generation of new ideas could occur. There was no intent to reach agreement or develop group recommendations. The issues under discussion, which were determined by a series of interviews, were the following: *a)* criteria for the designation of both occupied and unoccupied habitat; *b)* designation process; *c)* potential for exclusions from designation, *d)* economic impacts of designation, and the evaluation of those impacts; *e)* communication, incentives and partnering approaches.

The case shows that national workshops dealing with conservation policies can also identify distributive impacts of the policy (*e.g.* the economic impacts of designation). The participation of stakeholders in the process might make the policy more acceptable to the different affected parties.

6.4. Resolving conflicts in biodiversity policies

For groups likely to lose out from biodiversity policies, the potential for conflict is real. Conflict resolution can help manage conflicts induced by biodiversity policies and can also handle distributive issues in biodiversity policies.

Conflict resolution is closely linked to the participatory methods discussed in the previous section. If the policy design does not include negotiations with stakeholders, or there is no forum for the affected groups to express their interests and concerns, conflicts are likely to occur. Avoiding such conflicts requires some understanding of the motivations of affected social and economic groups. Income loss from the designation of a protected area, for example, will clearly cause resistance to the protected area and can easily be strong enough to reverse the designation. Reduction in access and loss of non-monetary benefits can also compromise the policy agenda, when sufficiently widespread. These responses may be as general as civil disobedience campaigns, or as focused as legal proceedings challenging the loss of income through lost property rights.

In practice, conflicts have also arisen through smaller-scale changes, such as:

- Designation of a new protected area.
- The preparation of a new management plan, or the revision of an existing one.
- Introduction of zoning systems.
- New regulatory regimes for natural areas.
- Reintroduction of protected animals, *e.g.* predatory species that can cause harm to local landowners and users.
- New fiscal measures such as taxes, or transferable quotas.

Most of the time these conflicts are closely linked to potential or perceived distributive effects of the policy measure. Table 6.7 gives some examples of potential conflict situations.

Conflicts related to biodiversity policy issues are often quite complex. There are usually many stakeholders with different interests, making

Table 6.7. **Examples of potential conflict situations**

Examples of biodiversity policy measures	Potential affected groups	Perceived (distributive) problem by the group	Potential trigger of conflict
Designation of a new protected area Preparation or the revision of a management plan Introduction of a zoning system	Local users Local land owners Native groups	Access is limited or costs are higher than before Use is limited, foregone income	No consultation or only official consultation at a late stage No compensation
Reintroduction of protected animal species (*e.g.* wolves, bears, seals)	Local landowners Local users of resources (*e.g.* hunters, fishermen)	Damage done by protected animals	No consultation and no information No compensation
Introduction of new taxes, user fees Introduction of transferable quotas	Users of natural resources	Unfair distribution of costs and benefits Foregone income	No consultation or only official consultation at a late stage

negotiations difficult (*e.g.* many landowners and users in a given area). Some stakeholders may not be as organised as others, so particular perspectives might not be represented in a unified manner, thus undermining the unity of purpose and power they would otherwise have. Some impacts appear only in the long term, and the effects of a particular biodiversity policy measure or activity might be not known in advance or there might be scientific debate about it. The policy measures may have asymmetric impacts, even within local communities – some may gain while others lose.

Torrell (1993) notes that the traditional government approach to conflict offers little or no resolution at all. Procedures tend to involve official answers to letters, possible examination of problems at different administrative levels (local authority, chief authority), and in worst cases, public hearings or court action. These procedures often lack personal contact with the other parties, or if there is any it is very official and superficial and does not help reach mutually acceptable solutions.

Research shows that consultations and alternative dispute resolution techniques (Box 6.3) often bring better and more satisfactory results than formal and official procedures like court cases.

These conflict resolution procedures are alternatives to administrative procedures and court cases and have many advantages (Torrell, 1993):

- **Sustainability of outcomes:** often these alternative conflict resolution procedures result in better and longer-lasting decisions because they satisfy the needs of all parties. The negotiation process and the outcomes are controlled by the groups involved.
- **Better climate for resolution:** the process is usually voluntary and with a good facilitator/mediator the personal conflicts can be reduced to a

PEOPLE AND BIODIVERSITY POLICIES – ISBN 978-92-64-03431-0 – © OECD 2008

Box 6.3. **Some alternative dispute resolution techniques**

- **Unassisted negotiation:** if the topic under discussion is not particularly complex, and there are not many parties to the conflict, negotiations can be carried out without external help. Well structured agendas and enough time to build trust among participants will be the most important conditions for success.
- **Facilitation/mediation:** this is a form of negotiation with the assistance of an impartial and skilled person with no stake in the issues under dispute. Negotiations are often difficult to organise and conduct successfully. As a result, mediators increasingly have been called upon to help parties convene negotiations, to prevent impasse during the negotiations, or to assist parties to continue when their discussions reach an impasse. The mediator helps the parties to improve communication, analyse the conflict, identify interests and explore possibilities for mutually agreeable solutions. Sometimes facilitation is distinguished from mediation. In the latter there is a greater role for the independent helper in guiding the participants.
- **Mini trial:** mini trials are commonly used to resolve conflicts outside the court. Parties are usually represented by a high official with the authority to agree with the decision. First, the principals in each party generally attend personally. Second, attorneys for each side are given an agreed amount of time to present their best arguments before a private neutral and the principals. Third, the mini trial is conducted by a neutral person agreed upon by all sides. After the presentations are completed, the principals meet privately in an attempt to settle the matter, with the neutral sometimes shifting roles from judge to mediator.
- **Arbitration:** in contrast to mediation, arbitrators conduct hearings and issue an opinion, either binding or non-binding, by advance agreement of the parties. Arbitration is often considered when the legal issues are not in dispute, but what is being contested is their application to the different factual circumstances of the case.

Source: O'Leary and Bingham, 2004, adapted with some modifications from *www.resolv.org/tools*.

minimum. Participants will be more open if they see that their needs are considered important.

- **Cost effectiveness:** usually the procedure is shorter and the costs are usually much lower than a court case.

However there might also be obstacles to using these methods in the public administration. These are listed below, along with probable solutions to overcome them (Torell, 1993):

- **Procedural compatibility:** traditional agency culture favours administrative procedures. Court-based and official administrative dispute resolutions are embedded in public administration in many countries. They are also supported by a regulatory framework. Training of ministry and agency personnel in dispute resolution techniques might help change this culture and attitudes. Successful cases show the effectiveness of these methods.
- **Lack of authority:** there is usually a lack of incentives for and authority of the participants to settle disputes in existing planning approaches. Existing planning approaches promote positional bargaining. Agency personnel and interest group representatives sometimes negotiate with a given mandate and do not hold authority to make decisions and settle disputes. Changing the human resource policies in the agencies with the introduction of a more open and collaborative working climate, provision of guidelines on bargaining and training of personnel can help overcome this obstacle.
- **Lack of awareness:** administrative agencies are frequently not aware of these alternative conflict resolution techniques. This can be addressed through training, and providing guidelines and a summary of successful cases.
- **Prevailing misperceptions:** there might be a feeling in the public administration that these alternative techniques produce a much weaker solution. However, in reality the success rate is very high (*e.g.* in the US: Bingham, 1986); information on best practices and education can change this misperception.

The above discussion shows that introducing these alternative dispute resolution techniques in public administration and land use planning may be worthwhile. However, issues of procedural compatibility need to be resolved and resources to build capacity in using these techniques need to be set aside. Additionally, training in and awareness about these techniques need to be initiated.

6.4.1. The role of government agencies in conflict resolution

When a conflict occurs it is important that the ministry or government agency in charge of biodiversity policy design and implementation is prepared for conflict resolution. The following steps are suggested and each discussed below:

1. Prepare for the negotiations.
2. Negotiate with the affected parties/stakeholders.
3. Implement the agreement reached during the negotiations.

PEOPLE AND BIODIVERSITY POLICIES – ISBN 978-92-64-03431-0 – © OECD 2008

STEP 1: *Preparing for negotiations (analysis of the conflict situation)*

Stakeholder analysis:* It is always useful to identify the main stakeholders in a conflict, explore the relationships between them, their attitude towards the situation and each other. There are a few tools that can be used:

- **Stakeholder characteristics matrix:** drawing up a simple stakeholder matrix can clarify the different affected groups, their main characteristics and their interests in the conflict.
- **Stakeholder-stakeholder relations and conflict matrix:** shows how the different stakeholders relate to each other; whether there is personal, structural or information conflict; or whether they have different values or interests in the situation. A chart can be used to group the stakeholders with similar interests.

After the identification and characterisation of the stakeholders, it is helpful to analyse the conflict cases and try to understand the reasons behind them (Moore, 1996):

- **Conflicts in personal relations:** it is common for parties in a conflict to have difficulty inter-relating at a personal level. This is sometimes hidden in emotions, misperceptions, misunderstanding, miscommunication, or the valuation of repetitive negative actions by the other parties. In some countries, this occurs in a context where nature conservation policy is seen as authoritarian and threatening. Public officials may thus be viewed as "enemies", not ready to engage in dialogue. Such a backdrop will magnify professional and personal conflicts and make it difficult to resolve the issues in conflict.
- **Information problems:** as emphasised in Part II, often the parties in a conflict have asymmetric information, or they interpret the same pieces of information in a different way. Scientific information and data, or legal texts, might not be easily understood by non-expert groups and these problems can make it more difficult to reach agreement. In biodiversity policy planning sometimes not enough information is released about the proposed measures, or not in time.
- **Structural problems:** the biodiversity policy-making process usually operates under time constraints, but needs to solve complex problems. Power may be distributed unevenly across stakeholders, irrespective of factors such as constituency size, or relative economic impact. In addition geographical or other physical circumstances can make the process longer.

* For further reading see: Grimble *et al.*, 1995; Start and Hovland, 2004; Herrero and da Passano, 2006.

Conflicts stemming from structural problems can also occur when governmental policies are not fully coherent, *i.e.* they introduce contradictory incentives. For example, intensive agriculture or forestry might be subsidised by the government while agri-environmental measures are also supported financially.

- **Different values:** heterogeneity in stakeholder groups will create different values – some of which are easily monetised, while others are not. Values are inherent in cultural, religious and other aspects of stakeholder groups and will change only very slowly. It can thus be a challenge to find sufficiently common values to begin a discussion in which trade-offs can be made.
- **Different interests:** interests should, of course, be the main focus of dispute resolution: personal, informational, and structural problems need to be solved before clearing the way for discussions on interests. These interests will have hierarchies which should be dealt with strategically so that opportunities are created for trade-offs. The most powerful interests are basic human needs like security, economic well-being, a sense of belonging or recognition. If a biodiversity policy proposal interferes with one of these, policy-makers are likely to face strong opposition from the affected economic or social group.

For biodiversity policies, dealing with interests represented by each stakeholder group engages the core issues causing conflict: the redistributive elements of a proposed policy. Understanding the source of each group's interests, how they might be affected and what can be done to mitigate that impact will be key to resolving any conflict. Calculating the costs and benefits of the proposed policy measure and their distribution among the stakeholders is thus a good basis for the negotiations. Tools like the ones shown in Part I can be used: extended CBA with distributional matrices; measures of equality of income distribution (*e.g.* Lorenz curve); multi-criteria analysis; social accounting matrix; or employment based analysis. If the conflict is over land use it can be helped by a resource use map, which is a very good tool for showing the affected area and the conflicting use patterns.

STEP 2: *Negotiations*

There are many possible outcomes of a particular conflict. If the negotiating partners think only in terms of winners and losers, no mutual gains will be possible. Win-win solutions satisfy both parties in the negotiations, but since the gain from win-win is smaller than the gain from win-lose, only circumstances that induce collaboration will achieve that result.

PEOPLE AND BIODIVERSITY POLICIES – ISBN 978-92-64-03431-0 – © OECD 2008

Negotiation and bargaining theorists distinguish two types of bargaining techniques: distributive/positional bargaining and integrative/principled bargaining. In a distributive/positional bargaining one actor demands a reallocation of resources from another actor, which the other actor opposes. They tend to ignore the problem addressed, but concentrate only on their own interests and see the others as adversaries. In their view it is a zero sum game, where one can win only at the expense of others.

In an integrative/principled bargaining actors try to find mutually beneficial outcomes. This is characterised by a search for innovative and creative win-win solutions. Negotiations are co-operative and actors see each other as partners. This is a positive sum game (Fisher *et al.*, 1991; Humphreys, 2001). Table 6.8 shows the main differences between the two approaches.

Table 6.8. **Main differences between distributive/positional and integrative/ principled bargaining**

Distributive/positional bargaining	Integrative/principled bargaining
Participants see each other as enemies	Participants trust each other
There is a power difference between the parties	Participants hold equal power
The goal is the victory (they are thinking in win-lose terms)	The goal is to reach agreement (they are thinking in possible win-win solutions)
Focus is on bargaining	Focus is on co-operation
Focus is on positions	Focus is on interests
Short-term individual benefits are important	Long-term mutual benefits are important
They see only one solution and stick to it	They see a range of solutions and they are able to discuss them
Good atmosphere is not important	Good atmosphere is important

Negotiation theorists emphasise that in conflict situations it is important to move towards integrative/principled bargaining. Therefore the principles of successful negotiations are as follows (Fisher *et al.*, 1991; *www.resolv.org/tools*):

- Understand the role of interpersonal dynamics in negotiations and help people move on.
- Discuss and address interests.
- Generate a wide range of options, minimising judgements at first.
- Agree on criteria by which to judge options for resolution.

STEP 3: *Implementing the agreement*

Once agreement is reached, follow-up and monitoring become necessary. If an agreement is "self-enforcing", then minimal follow-up arrangements are needed. On the other hand, difficult negotiations over easily-violated terms of agreement will call for more extensive follow-up arrangements. The regime for implementation and monitoring may have to be an integral part of the negotiation outcome.

6.4.2. *Conflict and biodiversity policies: some examples*

In this section we review five selected cases of conflict (summarised in Table 6.9) that highlight the key issues covered above. These case studies illustrate the general nature of conflicts around biodiversity policies. They also show the variety of possible stakeholders drawn into the resolution process and the broad range of possible areas of conflict. Most importantly, perhaps, the case studies demonstrate the contributions made by conflict resolution techniques in settling the conflict between the main stakeholders.

Opposition to the designation of protected areas in Germany (Stoll-Kleemann, 2001)

This case is one of our motivating examples in Part I of this volume (see Section 1.1.2). Here disputes arose around the designation of protected areas, with authorities as the main initiators. Opposition was expressed by local and regional authorities (*e.g.* mayors), forest administrators, local farmers, landowners, the hotel and holiday industry, local communities and people using the area for recreation. The conflict manifested itself in boycotts of public meetings on the establishment of protected areas, and public demonstrations and campaigns.

The example illustrates how the designation of protected areas can be reversed if conflicts are not solved in the planning process. In this case all elements of a conflict can be depicted: personal, information, structural, value and interests. The main lesson learned is that introducing designation by administrative decree is insufficient in the presence of a sizeable and well-defined group of policy losers. In such cases, more participatory methods are needed in order to prevent such conflicts, reveal the perceived distributive issues, and develop solutions.

Ria Lagartos Biosphere Reserve, Yucatán, Mexico (Fraga, 2006)

The Ria Lagartos Biosphere Reserve is one of the top 10 priority reserve areas in Mexico, receiving financing from the World Bank and other development institutions. There are four villages in the reserve, whose total population ranges from 800 to 2 500 inhabitants. Within the reserve, decision-makers and administrators have focused on biological conservation, failing to understand local social and political issues. Local people were not involved in planning and management and were not informed that they lived in a protected area; they only realised this when restrictions were imposed on

 PEOPLE AND BIODIVERSITY POLICIES – ISBN 978-92-64-03431-0 – © OECD 2008

Table 6.9. **Synopsis of selected conflict cases**

Case	Main stakeholders affected by the policy	Main points in the conflict (inc. distributive issues)	Resolution techniques used and some results
Opposition to the designation of protected areas, Germany	Local and regional authorities, forest administrators, local farmers, landowners, hotel and holiday industry, local communities, tourists	Designation is perceived as restricting the use of the area No stakeholder involvement in the process Negative attitude towards nature conservationists Boycotts of public hearings, many designation processes have failed	No resolution techniques (But author proposes: more participatory methods, landscape preservation association with all stakeholders represented, use of facilitator)
Conflicts because of restrictions in a biosphere reserve, Mexico	Local communities, local industry	Zoning and restrictions on the use of the area (*e.g.* salt industry) No involvement of locals in the planning process Conflict because of restrictions and zoning	Efforts to involve locals in conservation activities Training of locals in resource management Public forums for the revision of the management plan Establishment of the Reserve Technical Advisory Committee (Mixed results)
Conflict around reindeer breeding and hunting, Sweden	Saami people (indigenous group)	Permits for amateur small game hunting and fishing on land originally designated to Saami people for reindeer breeding	European Court of Justice Commission on Hunting and Fishing appointed to clarify the scope of Saami rights
Diseased bison in Wood Buffalo National Park, Canada	Indigenous people	Plan to kill infected bison Plan opposed by Aboriginal groups	Consultation with native people: new plan Management board set up
Reintroduction of Mexican wolf in Arizona and New Mexico, USA	Livestock industry, native tribes	Livestock killing, danger to tribes Poor communication	Three-year review workshop involving main stakeholders: problem identification and formulation of recommendations Five-year review of the programme Moratorium on releasing more wolves above a certain population size

them for using and accessing natural coastal resources (*e.g.* cutting wood). The conflict arose in the late 1980s and early 90s when zoning and restrictions were introduced governing the expansion of the salt industry and prohibiting certain traditional resource exploitation activities within the reserve. The first management plan was approved without consultation with the locals. The policy has slowly changed. Starting from the mid-1990s, public forums have been held among local users, social organisations and academic institutions to revise the first management plan.

Recently there have been efforts to involve the community in conservation activities. The results are mixed so far, but the first steps have been taken. Local people have been trained in resource management under the United Nations Development Programme, but this is not perceived as useful by the participants. There is still a lack of dialogue between holders of local knowledge and those with scientific knowledge. The administrators have also seen failures, which in their view resulted from difficulties of local users in administrative work, their ignorance of aquaculture management, the staff restrictions placed on projects by the reserve and lack of internal organisation among the users.

Dialogue is now occurring among the parties through the Reserve Technical Advisory Committee. Although its objectives are well planned, the methodologies for carrying them out have not changed. The effective integration of some local communities is still not occurring because the communities do not see any incentive for taking part. Thus, the same people are involved all the time and the diversity of the community is not represented in the dialogue.

This case shows that conflicts can arise if zoning and restrictions are introduced without involving users in the planning process. There is also a distributive issue because former users are restricted in their activities (potential loss of income). The attempts to solve these conflicts involve a learning process: how to involve locals in biodiversity policy planning and managing resources, and how to establish a more formal decision-making body. The process is long and needs revision from time to time as new conflicts arise (*e.g.* between local and scientific knowledge, locals' attitudes to different tasks and problems of community representation in decision-making).

The Saami people in Sweden (adapted from OECD, 2004)

Sweden's Reindeer Husbandry Act (1971, last revised in 1993) allows the reindeer-breeding Saami some autonomy over their own affairs. They are permitted to herd and they enjoy special land and water rights. But the legislation does not give such rights to Saami who live by fishing or other occupations. Moreover, since the act's passage, the reindeer breeders have lost large tracts of pasture to clear-cutting and ploughing.

In 1993, parliament established the Saami agency, but at the same time amended the law to permit amateur small-game hunting and fishing in the reindeer-grazing mountains of Jämtland, and west of the cultivation boundary in Västerbotten and Norrbotten. The Reindeer Husbandry Act had originally designated these lands exclusively for reindeer herding all year round. With this amendment, the Saami's exclusive rights were revoked in favour of

PEOPLE AND BIODIVERSITY POLICIES – ISBN 978-92-64-03431-0 – © OECD 2008

parallel hunting rights on these lands. Even before this, responsibility for hunting permits had largely been taken over by county authorities, and permits were granted to non-Saami people upon payments to the Saami.

The change of policy on hunting was opposed by the public and by legal and environmental experts; the dispute has now been brought to the European Court of Justice. It is still on the government's agenda, along with questions about the rights to land and water in the Saami area, which are being reviewed by a committee appointed in 1998.

A Commission on Hunting and Fishing was appointed in April 2003 in order to clarify the scope of Saami hunting and fishing rights, and to propose more precise regulations by December 2005.

This is a good example of what happens if a policy changes the rights of indigenous people without prior consultation. It also shows that distributive issues can have both economic (loss of reindeer herding) and social (violation of historical rights) elements. If such conflicts are not prevented or solved through extensive dialogue, they can end in a court case. The early use of a more participatory method (like mediation/facilitation or a mini trial) might have led to a more expeditious process, and might have had a more satisfactory result.

Diseased bison in Wood Buffalo National Park, Canada (Nepal, 2000)

There is archaeological evidence that indigenous people have inhabited the Wood Buffalo Region of Canada for more than 8 000 years. The Wood Buffalo National Park has a long-standing tradition of native subsistence use by indigenous groups, including hunting, trapping, fishing and the seasonal collection of edible plants and berries.

In 1989, 30 to 50% of the North American bison (*Bison bison*) (a protected species) in the park were reported to be infected with bovine brucellosis and tuberculosis. An assessment panel recommended that all diseased bison be culled and replaced with bison from another national park. The plan was opposed by Aboriginal groups, environmentalists and other concerned organisations, and local citizens. This opposition resulted in a new plan to "test and slaughter" infected bison, formulated after consultation between the federal departments of environment and agriculture and local native people. According to the new procedure, bison would be rounded up, tested for the diseases, and only slaughtered if tested positive.

A management board was set up (the Northern Buffalo Management Board) which operated for one and a half years. It was made up of federal, territorial and Aboriginal community representatives who worked together to try and produce a consensual approach to dealing with some of the park's bison health problems. Data collection used both traditional knowledge and scientific methods. Aboriginal communities were involved, and, with funding, developed their own plans for handling the disease. This was a good example of temporary co-management.

The case shows that discussions with affected groups (native people) can help find mutually acceptable solutions even in serious situations where immediate action is needed. Setting up joint management bodies is also one possible way to address future problems (and distributive issues as well).

The reintroduction of the Mexican wolf in Arizona and New Mexico, USA (Kelly et al.*, 2001, Unsworth* et al.*, 2005)*

The Mexican grey wolf (*Canis lupus baileyi*), has been gradually reintroduced into Arizona and New Mexico since 1998. In the first year, three family groups (11 wolves) were released and the aim was to increase the population to 100 wolves over 1.2 million hectares. The so-called Blue Range reintroduction project is managed jointly by the Arizona Game and Fish Department, New Mexico Department of Game and Fish, USDA Forest Service, USDA-APHIS Wildlife Services, White Mountain Apache Tribe, and the US Fish and Wildlife Service. These organisations form the Mexican Wolf Adaptive Management Oversight Committee (AMOC) (*www.fws.gov/southwest/es/mexicanwolf/BRWRP_home.shtml*).

In terms of expanding the wolf population, the programme has been successful (there were 44 individuals in 2004), but there have been some conflicts with native tribes and cattle farmers. The latter were losing livestock to wolf attacks. They reported their loss for compensation and tried to ban new wolf releases above a certain population size. While tribal areas were outside the project areas, wolves started to appear in native land. Some tribes asked for the removal of wolves from their land (San Carlos Apache Tribe), but others (White Mountain Apache Tribe) signed a co-operative agreement with the US Fish and Wildlife Service to allow wolves into their area (*www.fws.gov/southwest/es/mexicanwolf/chronology.shtml*).

In 2001 a three-year review was conducted, involving a workshop with the main stakeholders. The participants summarised the main problems and formulated some recommendations. Problems included: *i)* inadequate mechanisms for communicating with stakeholders; *ii)* conflict between rural and urban values, perceptions and points of view; *iii)* actual losses to individuals and local communities that are not being adequately addressed; *iv)* better consideration of full costs of the programme needed (Kelly *et al.*, 2001).These comments show that there are still social and economic problems to be solved.

A five-year programme review was also compiled, including a socio-economic analysis. It is estimated that five to 33 cattle are killed by wolves

PEOPLE AND BIODIVERSITY POLICIES – ISBN 978-92-64-03431-0 – © OECD 2008

every year, which is less than 1% of the number of cattle grazed in the area. The number of other animals killed (sheep, horses, dogs) is lower. According to the report, the total value of lost livestock to ranchers was estimated at between USD 38 600 and USD 206 000 (1998-2004). Since 1998, USD 34 000 in compensation has been paid to ranchers. There are two Apache tribes who have land adjacent to the main wolf area. Although each tribe initially objected to the introduction of wolves onto their land, now one of them has an agreement with the managing body to allow wolf introductions. The other tribe still objects and is complaining about uncompensated losses of calves (Unsworth *et al.*, 2005).

In 2005 AMOC approved a moratorium for 2006, stating that it would not allow new releases of the Mexican wolves in 2006 if the number of breeding pairs in the wild was six or more on 31 December 2005. The decision was made after consulting with members of the livestock industry. This shows that there is still opposition to the reintroduction programme and adaptive measures are needed (*www.fws.gov/southwest/es/mexicanwolf/documents.shtml*).

This case is a good example of the complexities of reintroducing a protected predator species and the need for managing evolving conflict situations. Even though there is a stakeholder committee that oversees the project, opposition still occurs from time to time, requiring the review of the project and decisions on certain actions.

In the next chapters we will show how involving local economic and social groups can prevent conflicts over natural resources. We will also outline the requirements for effective and mutually satisfactory co-operation.

ISBN 978-92-64-03431-0
People and Biodiversity Policies
Impacts, Issues and Strategies for Policy Action

Chapter 7

Institutional Approaches: Property Rights, Compensation and Benefit-sharing

7.1. Introduction

Biodiversity policies commonly involve changes in how land is used. For previous users, this means additional costs, lower benefits or a foregone profit. Institutional changes are a way of responding to these changes in the welfare position of those affected by the policy. They do so by granting them a degree of control in the form of a property right, use right, or entitlement or by offering them participation in newly created institutions such as contract schemes or new markets.

A typical example of an institutional change[1] is the right to compensation for the extra costs incurred, thus mitigating the distributive effects of the policy. In other cases, voluntary programmes are designed by government agencies to attract private landowners and users to participate in newly created markets for conservation easements or in novel contracting mechanisms.

The literature on novel institutional instruments in biodiversity policy-making shows that they have specific distributive impacts. In the context of market-based systems, Pagiola *et al.* (2005) showed that PES (payments for environmental services, see Section 3.2.2) can help reduce poverty. PES combine the creation of property rights with the delivery of environmental services (such as clean water), as well as introducing a market where these services can be sold or bartered. Examining the potential impacts using a social accounting matrix and estimating the relative size of gains and losses on the basis of field experience in Latin America, the authors found that PES are generally pro-poor. Important qualifications concern the security of property rights and the extent to which the creation of property rights and their marketability induce changes in the production of environmental services. If – as a result – capital is substituted for labour in the production, there may be second order impacts on the local labour market, in particular for farm workers. These groups may in fact lose income as a result of PES being introduced.

A characteristic of PES is therefore that they turn the resident population into providers of ecosystem services by creating titles through markets. This change of ownership has distributional consequences. Zbinden and Lee (2005) showed that in Costa Rica – where payments for environmental services were introduced in 1997 – there is a clear pattern to benefits. Those with an above

PEOPLE AND BIODIVERSITY POLICIES – ISBN 978-92-64-03431-0 – © OECD 2008

average endowment of assets (agricultural land or forests) participate most and reap the highest benefits from these markets.

Experiences of other institutional changes are documented by Adhikari (2005) and Lybbert *et al.* (2002). In the case of the devolution of property rights from governments to resident communities, Adhikari found an inverted U-shape of dependency on forest and income in communally managed forests – going from low dependency, to high, and back to low as income increases. Property rights devolution to communities can therefore have surprising effects on the distribution of benefits. Lybbert *et al.* (2002) examined the distributional effects of market creation for a commodity (plant-based argan oil) whose discovery and continued production is intimately connected with biodiversity maintenance (land-use change is responsible for its loss). Theoretically in this case, argan oil could successfully combine market-based conservation with local benefits. Empirically, however, this idea is not borne out. There is no measurable impact of argan oil commercialisation on local development and poverty reduction. Instead, the distribution of benefits across households within the population appears to be regressive, both regionally and between households. This points to better access to markets by the better-off (Lybbert *et al.*, 2002).

The main focus of this chapter is on the most popular institutional policy tools for resolving distributional inequities: schemes which compensate for restrictions and voluntary compensation in the form of contracts. These tools put the government in the driver's seat, using funds raised through general taxation as transfer payments to affected parties. However, this chapter also covers institutional changes, such as conservation markets in which the government has a hand in creating the market, but then assumes only a supporting role (*e.g.* in the form of subsidies to market transactions), or no active role at all. Examples illustrate the successful operation of these programmes.

7.2. Main features of compensation schemes and voluntary agreements

Nature conservation regulations often affect the activities of private landowners and users by limiting the use of their land or obliging them to carry out certain conservation activities. These obligations can mean an opportunity cost and/or some direct cost to the owner or user. To some extent these costs are acceptable at the social level, but they might involve too large a burden on the landowner, at which point the distributional effects need to be dealt with. Compensation[2] and voluntary agreements are two ways of inducing private owners and businesses to carry out biodiversity measures.

Compensation schemes for involuntary restrictions and voluntary agreements have been used particularly frequently where private ownership and private use are dominant. These include protected areas, buffer zones of protected areas, environmentally sensitive areas or valuable natural/semi-natural landscapes. In some areas natural values are conserved through nature-friendly agricultural management, sustainable forestry or ecotourism activities. In other regions, private protected areas are becoming important for biodiversity conversation (Langholz and Krug, 2004). There are also some biodiversity-rich private properties whose owners do not cultivate their land, using it for family recreation or for other small-scale activities.

Voluntary schemes generally involve collaborative approaches, where agreement is reached between the government and private parties. Most often they mean that the owners or users do more than is obliged by law. Distributive issues are generally settled through negotiation: in some cases costs are not compensated, but accepted (there may be expectations of non-monetary subsequent benefits or averted costs). The schemes can be introduced after a careful planning process, but after the introduction there is no real negotiation with the landowners/users. In other cases they can include intensive negotiations and discussion about the requirements and about the terms of the contract.

Table 7.1 compares the main characteristics of the two approaches. Some of the examples are discussed in more detail in Sections 7.2.1 and 7.2.2 below.

Positive incentives can be used to attract private owners and businesses into the nature conservation programme. These may include:

- Restricting further regulation (no additional regulatory burdens). In some cases private actors are willing to undertake conservation actions just to avoid further regulation. They enter into an agreement with the authorities, prepare a management plan with assistance and accept certain restrictions and sometimes restoration obligations in their property. As an incentive the promise is made that no additional regulatory burdens will be imposed. The Safe Harbour Program in the USA is a good example (see below).
- Technical assistance. This can be an important complement to voluntary programmes. Landowners and users often are not expert in conservation measures, habitat and species protection, rehabilitation, nature friendly farming or forestry management. With some assistance traditional methods can be reintroduced and combined with modern techniques. Examples include technical assistance to farmers participating in the Entry Level Stewardship programme in UK, or advice to landholders participating in the BushTender Programme in Australia.
- Financial incentives (*e.g.* payment, conservation banking credits, tax reductions). There is considerable variation in financial incentives used to

Table 7.1. **Main characteristics of compensation schemes and voluntary agreements**

Characteristics	Compensation schemes for involuntary restrictions	Voluntary agreements
Main characteristics	Restriction is imposed on the landowner or user by regulation Accepting the restrictions and fulfilling conservation obligations are not voluntary Compensation is a joining measure	Landowner or user takes voluntary biodiversity measures and accepts restrictions and other requirements Contract/agreement is signed between the land user and the authority There are usually joining financial incentives as well
Strengths	Easier to manage and introduce. There are usually a few compensation categories (*e.g.* based on the types of land) and basic calculations	Co-operative approach Possibility of mutual gains A learning process occurs Creative solutions possible The landholder's autonomy is respected A voluntary activity Participants do more than required by law
Weaknesses	Participation is not voluntary There is no incentive to do more than required by law It is not tailored to individual situations Cost calculations might be either under- or overestimated	Requires much more participation by government and state administration (during both preparation and monitoring)
Redistribution effectiveness	Costs are compensated	Negotiations or a careful planning process mean the solution is acceptable to participants: costs and profits foregone often compensated for, or other types of financial and technical incentives available
Most appropriate policy situations	For large homogenous areas and simpler management requirements When the costs of the scheme are large enough to require significant inducement	Where the cost of accepting an agreement is small, or where the threat of more costly measures is credible
Examples	Compensation for restrictions on Natura 2000 sites, European Union Compensation during the establishment of Neusiedler See-Seewinkel National Park, Austria	Safe Harbour Program and other voluntary schemes, USA Habitat Stewardship Program and agri-environmental Greencover Program, Canada Natural Forest Reserve Programme, Austria Agri-environmental scheme, European Union, (*e.g.* Entry Level Scheme, UK) Voluntary scheme for forest protection (METSO programme), Finland BushTender programme, Australia

encourage private landowners and businesses to carry out conservation activities. These are different from the compensation schemes mentioned earlier because they change marginal prices. They also deal with distributive issues since they provide payment for actions. Table 7.2 gives an overview of some options, some of which are discussed in more detail in Section 7.2.1 below.

Table 7.2. **Overview of options in financial incentive schemes**

Type	Details	Examples
Grants for conservation developments	Yearly announcement of grants for which landowners need to apply (one-off payment or payment for the duration of the project)	USA: Grassland Reserve Program, Wetland Reserve Program – restoration agreements Canada: Habitat Stewardship Program EU: LIFE Nature
Yearly payment for carrying out conservation activities	Different schemes for the content of the payment: profits foregone, extra costs, incentive payments Different schemes to calculate the costs: given by the government, negotiated or bidding process Time-scale also varies	EU agri-environmental scheme BushTender, Australia USA: Conservation Reserve Program, Grassland Reserve Program, Wetland Reserve Program, Habitat Incentive Program
Conservation/mitigation banking credits	Credits are received for conservation measures that can be sold to developers in other areas	USA: wetland banks, grassland banks and other conservation banks
Tax reduction for conservation activities	For set-aside and for donation of land for conservation purposes: one-off tax reduction or credit	Canada: Ecological Gifts Program

In some cases grants are provided to cover the total or part of the costs of recovery or rehabilitation measures (cost-sharing). These are either a one-off payment or a series of payments over the life of the programme. Examples can be found in the USA, in Canada and in Europe.

In other cases, yearly compensation is given to the farmers or other land users for profits foregone or to cover additional costs. Sometimes even extra payment is provided to encourage them to join a programme. This is the case for the European agri-environmental scheme in most European countries and the BushTender programme in Australia. In this latter programme the payment is given through a bidding process and funds are allocated according to the "best value for money".

Conservation banking credits are an innovative financial incentive: they mean that landowners enrolling in the programme (and agreeing to save a certain habitat or species) get mitigation credits, which they can sell to other landowners who need to mitigate their land development impacts on listed species (US Department of Interior *et al.*, 2005b). They are very popular in the USA, where wetlands and other biodiversity-rich areas are used as a basis for banking.

Tax reduction for conservation efforts is another innovative way to attract landowners and farmers into biodiversity related activities. In Canada's Ecological Gifts Program, landowners who donate their land or partial

 PEOPLE AND BIODIVERSITY POLICIES – ISBN 978-92-64-03431-0 – © OECD 2008

interests in their land to qualified recipients (governmental agencies, certain NGOs) are able to receive tax reductions.

The government can enhance these incentive programmes in several ways, including by:

- Training governmental officials: as mentioned earlier, governmental officials are not always trained in techniques that induce and sustain co-operation.
- Creating a good clearinghouse mechanism: this can be a forum for sharing positive experiences, providing information about the various programmes, answering common questions and providing guidelines for joining the programmes. These have been working well in many countries, *e.g.* the USA and Canada.

Independent of whether the government plans a compensation scheme or a voluntary mechanism, distributive issues arise between the recipients of funds under such programmes and the general public from whom the funds have to be raised. In compensation schemes, more generous payments reduce the likelihood of conflict, but also mean that some recipients may receive more than would have been necessary to pay them to provide the same extent of participation voluntarily. In voluntary agreements, generous contract terms will generate high levels of participation, but the same policy outcome could in all likelihood have been accomplishable at lower aggregate cost. As the examples below show, various voluntary agreements incorporate features such as competitive tendering that attempt to use public funds as economically as possible in order to arrive at an acceptable trade-off between the interests of the taxpayer and the interests of the affected parties.

7.2.1. Examples of compensation schemes

Compensation on Natura 2000 sites across Europe (Council Regulation (EC) No. 1698/2005)

In 2003, the European Union amended its regulation on support for rural development from the European Agricultural Guidance and Guarantee Fund. The amendment allowed for payments to compensate farmers whose agricultural practices are restricted by the EU Bird and Habitat Directives, and who had incurred costs and foregone income. The maximum payment is EUR 200-500/ha, but this may be increased in justified cases, when specific problems arise (Council Regulation (EC) No. 1783/2003 amending Regulation (EC) 1257/1999). In 2005 a new fund was created, the European Agricultural Fund for Rural Development, which will serve as the main fund for the 2007-2013 EU budget period. It has taken over the previous compensation scheme, but extended it to forest areas. The maximum payment for agricultural areas is EUR 200-500/ha, and EUR 40-200/ha for forest areas (Council Regulation (EC) No. 1698/2005).

The compensation for the Natura 2000 sites is a way to mitigate the distributive effects of the EU's Birds and Habitat Directives.

Compensation during the establishment of the Neusiedler See-Seewinkel National Park in Austria (Hubacek and Bauer, 1999)

The Neusiedler See-Seewinkel National Park was established in 1983 and was the first national park in Austria to be recognised as an IUCN category II (natural monument or natural landmark). A reed bed within the park is recognised as a UNESCO biosphere reserve and the wetlands are recognised as internationally important under the Ramsar Convention. Prior to establishment of the park, landowners as well as other stakeholders used the land for agriculture, hunting, fishing, the reed industry and tourism. Compensation was provided to landowners who gave up land for the new park, legal hunters whose access had been restricted, and members of the fishing industry for ceasing to stock the lake with non-native fish species.

This compensation operated as an incentive for biodiversity conserving behaviour and to meet opportunity costs, and at least in the case of farmers, was determined through negotiation (Hubacek and Bauer, 1999). Financial penalties for disregarding park laws were also introduced. Extensive monitoring of the ecological state of the national park has been underway, but reports of the data from the relevant Austrian agencies are pending. At least one species, the ferruginous duck (*Aythya nyroca*), which had declined to effective extinction in the 1980s at Seewinkel, has re-colonised the site, suggesting that habitat improvements have occurred. In this example, negotiated compensation was used, implying that stakeholders were probably provided with adequate economic benefits to meet their opportunity costs. Negatives incentives were also used. If the state of habitats for other species is indeed improving, these approaches have been successful in meeting their biodiversity conservation objectives. Once biodiversity friendly management becomes routine and new markets have evolved, then compensation can be adjusted to actual economic losses.

7.2.2. Examples of voluntary schemes

As discussed in Part I, voluntary instruments are non-coercive, price-based systems. Their aim is to retain as much of the institutional structure as possible, but to change the rewards for individuals taking actions within the given institutional framework. Biodiversity policies using voluntary instruments increase the marginal net benefit of a conservation activity either by decreasing its costs or by increasing its gross benefits, leaving the extent of behavioural change to the participant. Voluntary schemes are therefore among the least invasive of instruments.

Here we present various examples of voluntary schemes in order to demonstrate how they can address distributive issues.

USA: Safe Harbor Program in combination with lease agreements (US Department of Interior et al.*, 2005a)*

The Safe Harbor concept was developed by Environmental Defense (a non-governmental organisation) and the US Fish and Wildlife Service to encourage private landowners to restore and maintain habitat for endangered species without fear of incurring additional regulatory restrictions. Between 1995 (when the first Safe Harbor agreement was signed) and March 2005, more than 325 landowners enrolled over 1.4 million hectares in 31 Safe Harbour agreements. Another 60 agreements are under development (US Department of Interior *et al.*, 2005a). Existing agreements are to be found in states across the country and benefit a variety of endangered species. The diverse group of landowners participating in Safe Harbor includes private forest owners, ranchers, residential property owners, corporate landowners, golf courses and a monastery (*www.environmentaldefense.org/article.cfm?ContentID=399*).

In many cases, Safe Harbor agreements are combined with conservation leasing, when the government pays the landowner for undertaking certain conservation activities. There is more than USD 100 000 of cost-sharing money available to restore habitat on private land. The administering body enters into a ten-year cost-sharing agreement with landowners in which they split the cost of habitat restoration activities, including prescribed burning, mechanical and chemical control of woody vegetation, fencing, and other activities. Technical assistance is also provided, which, from the landowner's perspective, adds significantly to the value of the programme (Environmental Defense, 2000).

The Safe Harbour Program is a good example of a voluntary programme which landowners join to avoid further regulation. Future distributive issues are settled immediately so it gives certainty to landowners. The cost-sharing programme offered by the state is an additional incentive and helps reduce any likely distributive effects.

The Habitat Stewardship Program of the Canadian Wildlife Service (www.ec.gc.ca/hsp-pih)

The Habitat Stewardship Program (HSP) for Species at Risk is a good example of voluntary conservation on private land, where distributive effects are reduced with the financial support of the government. Under Canada's Species at Risk Act, stewardship is the first step in protecting critical habitat.

The Canadian federal government approved CAD 45 million over five years for the HSP, beginning in 2000.

The programme fosters land and resource use practices that maintain the habitat necessary for the survival and recovery of species at risk, enhancing existing conservation activities and encouraging new ones. In its first two years, the programme established over 100 partnerships with Aboriginal organisations, landowners, resource users, nature trusts, provinces, the natural resource sector, community-based wildlife societies, educational institutions and conservation organisations. Stewardship projects resulting from these partnerships benefited the habitat of nearly 200 nationally-listed species at risk and well over 100 provincially-listed species at risk.

In addition to the above objectives, the programme aims to achieve 2:1 leverage on funds that it invests, so that for every CAD 1 provided by the HSP, CAD 2 is raised by project recipients, either by financial or in-kind resources (volunteered labour, products or services). Partner funding and other support broaden the scope of projects, improve on-the-ground results, and strengthen public and private collaboration (*www.cws-scf.ec.gc.ca/hsp-pih/*).

One example is the Appalachian Corridor Project, an initiative of the Ruiter Valley Land Trust in co-operation with the Nature Conservancy of Canada (Québec) and several other local organisations. This project aims to implement a trans-border conservation strategy. The goal is to protect the natural corridor that extends from the Green Mountains of Vermont through the Sutton Mountains range to Mount Orford in the Eastern Townships of Québec. Under the strategy, sites of significant environmental value are identified, then conservation plans are developed to ensure protection of natural environments, wildlife habitats, old-growth or exceptional forests, and at-risk animal and plant species. One of the world's largest concentrations of Bicknell's thrush (*Catharus bicknelli*), a species of special concern both in Canada and globally, is located in this region. Inventories have also revealed the presence of several rare or at-risk species, including plants such as American ginseng (*Panax quinquefolius*), an endangered species, and white wood aster (*Eurybia divaricata*), a threatened species.

Most of the Sutton Mountains region is privately owned, and its integrity is threatened by logging and real estate development. In 2001, the Ruiter Valley Land Trust carried out significant studies in the region aimed at identifying the habitats of species that are at risk or likely to become so. Individual conservation plans have now been developed for eight priority properties, indicating the ecological significance of the properties and suggesting, among other things, conservation measures to protect the species at risk and their habitats. These plans are helping to raise landowners' awareness and have provided a starting point for negotiating environmental

PEOPLE AND BIODIVERSITY POLICIES – ISBN 978-92-64-03431-0 – © OECD 2008

protection measures on up to 5 000 ha of private land. By the end of 2001, approximately 1 200 ha had received some level of environmental protection. The HSP contributed CAD 120 000 to this project, which has required a total investment of CAD 1.135 million (*www.cws-scf.ec.gc.ca/hsp-pih/*).

Greencover Canada: an agri-environmental programme (www4.agr.gc.ca/)

Greencover Canada is a five-year, CAD 110-million Government of Canada initiative to help agricultural landowners improve their grassland-management practices, protect water quality, reduce greenhouse gas emissions, and enhance biodiversity and wildlife habitat. It is a complex environmental-biodiversity programme for agricultural landowners with technical, and in some areas, financial assistance. Financial assistance is given for those programme components which put the largest financial burden on the participating farmer. Governments need to decide what level of negative distributive effects can be shouldered by the farmer (*i.e.* whether it is a socially acceptable requirement or whether positive market and other possibilities can compensate them), and what amount needs to be covered.

The programme focuses on four components: *i)* land conversion: converting environmentally sensitive land to perennial cover through technical assistance and financial incentives; *ii)* critical areas: managing agricultural land near water; *iii)* technical assistance: helping producers adopt beneficial management practices; and *iv)* shelterbelts: planting trees on agricultural land.

The critical areas and shelterbelt components are available through the National Farm Stewardship Program in each province. Eligibility for cost-sharing for Beneficial Management Practices (BMPs) available through these components includes the completion of an environmental farm plan. The technical assistance component provides financial support for organisations giving information to producers on Greencover BMPs.

Agri-environmental measures in the EU (European Commission, 2005)

The European Union's agri-environmental scheme (operating as part of the Common Agricultural Policy) is a good example of how the distributive effects of conservation activities can be mitigated by paying the farmer for additional costs and for loss of income.

The agri-environment measures encourage farmers to carry out environmental and biodiversity-related activities on their land (*e.g.* reducing pesticide and fertiliser use, using extensive farming methods, ensuring sustainable management of linear features like hedges and wildlife corridors, and implementing action to conserve local or threatened livestock breeds or plant varieties). Such payments are intended to reflect the broader value to society of these measures. Farmers sign a five-year contract with an official national administrative body and are paid for their additional costs and for their loss of income. In some circumstances, an incentive payment of up to 20% may be made. The schemes are different in every country, but the main aims, the types of measures and the maximum payments are provided by the EU. Member countries need to prepare a national rural development plan which details the country-specific goals, the proposed measures and the system of allocation. Agri-environmental payments are co-financed by the EU and the member states.

Evaluation studies show that the flexibility of agri-environmental measures enables policies to meet certain environmental needs which cannot be met otherwise. The great diversity in its implementation shows that it is able to respond to very diverse situations on the ground. The optional contractual nature of the measures makes it an instrument with a high level of acceptance among farmers, and a correspondingly high level of compliance. Agri-environmental measures play an educational role, they improve environmental awareness among farmers, and acceptance of farming practices among the general public.

The Entry Level Stewardship Scheme in the United Kingdom (summarised from Mowat, 2008)

The Entry Level Stewardship (ELS) scheme is a voluntary programme open to all UK farmers and land managers. It aims to continue to address three of the objectives of previous agri-environmental schemes (conservation of biodiversity, landscape, and the historic environment), and also adds natural resource protection as a fourth key objective. The previous programme had been targeted at land with the highest environmental value or potential value. The aim of the new programme is to encourage a large number of farmers to deliver simple yet effective environmental management beyond that of legislative requirements, across a wide area of farmland. The scheme provides a fixed payment per hectare in return for a package of management measures chosen by the farmer from a standard menu of options. Each option is worth a certain number of points per hectare (based upon the income foregone in carrying out the option), per metre or other appropriate unit, and the farmer must accumulate sufficient points to reach a threshold score proportional to the area of the farm.

The decision-making process prior to the introduction of the scheme involved many steps: consolidation of the evidence base (environmental and socio-economic aspects), three public consultations and a large scale pilot. Consolidation of the evidence base showed that agri-environmental measures had generated significant environmental benefits, were good value for money,

PEOPLE AND BIODIVERSITY POLICIES – ISBN 978-92-64-03431-0 – © OECD 2008

and were highly valued by both the agricultural sector and the general public. The analysis was supplemented with a consultation with the agreement holders, other stakeholders and staff involved in the administration. In addition a live pilot was run to evaluate the design. The pilot was launched in four areas of England in February 2004. A range of farming types (*e.g.* arable, lowland livestock) and geographical areas was represented. Success criteria were agreed in advance, covering uptake, delivery of environmental benefits, acceptability of the scheme to farmers, partners and the wider community and successful and efficient administration. Both the consultation and the pilot showed that there was strong support for the scheme.

The case illustrates the change in the operation of an agri-environmental measure to make it accessible to a larger number of farmers. It also shows that a combination of external reviews, consultation exercises and in-house investigation proved an effective way of evaluating the environmental benefits and economic efficiency of the existing agri-environmental schemes and of getting some information on how well the schemes operated and how they were perceived. The evaluation, combined with a pilot project, provided a good basis for the further development of the programme.

The Austrian Natural Forest Reserves Programme (Frank and Müller, 2003)

The Natural Forest Reserves Programme of Austria is a voluntary programme for private owners, in which distributive issues related to biodiversity friendly management are settled through a yearly payment.

This programme was initiated in Austria in 1995 to help systematically establish a representative network of natural forest reserves. A framework concept was prepared and was negotiated by a wide range of stakeholders. This formed the basis for the selection process and the management of the reserves. All other activities occur through close co-operation with individual forest owners and follow a bottom-up process. The following principles were negotiated and agreed: *i)* participation in the programme is strictly voluntary; *ii)* contracts are based on private law; *iii)* long-term commitments are mandatory: 20 year contracts, including the right of extension; *iv)* opting out is only allowed under defined circumstances; *v)* the compensation level reflects the income value the forest owner could have earned from his property, with an additional bonus. Since the programme started, 850 proposals have been submitted and 180 sites have been approved. The contract is based on an expert report and a survey, using a grid system of permanent sample plots for compensation assessment, which is suitable for future monitoring assessment of development. The programme has been very successful (*i.e.*, there have not been many violations or terminations of contracts).

Forest Biodiversity Programme in Finland (summarised from Horne and Naskali, 2006)

The Forest Biodiversity Programme for Southern Finland (METSO) was established in 2003. It was planned by the broadly-based METSO Committee with representatives from different authorities, NGOs and interest groups. The programme includes 17 projects aimed at preserving the biodiversity of forests in Southern Finland. All 17 programmes have already been launched and the first of these has already ended. The total assessment of the ecological, social, and economic impacts of METSO has been prepared and an English summary is available.[3] Among other instruments, the METSO programme contains two voluntary measures targeted at forest owners: *i)* natural values trading; and *ii)* competitive tendering.

Natural values trading means that the landowner agrees to maintain or improve specified biodiversity values on their forest holding; in return they receive a payment from the "purchaser" of these natural values (the state or the Forest Conservation Foundation). Since 2004, natural value trading has also been used by co-operative networks. The contract periods vary according to the site, but the longest is 13 years. Between 2003 and 2005, 93 contracts covering 871 ha were drawn up, with an average annual payment of EUR 162 ha. The most important output of this pilot has been the positive attitude by forest owners towards this conservation scheme.

Competitive tendering is another new instrument for the conservation of biodiversity, aimed at long-term or permanent protection of ecologically valuable sites. In competitive tendering, the forest owner can offer their forest either permanently or for some fixed-length period to the environmental authorities in return for compensation at a level proposed by the owner. Almost 40 proposals were made in 2004, representing a total of 800 hectares. Implementation by way of purchases, private protection or 20-year set-term contracts covered 9 areas, totalling 115.5 hectares, and costing EUR 500 000.

This case shows that there are many innovative ways to involve private forest owners voluntarily in nature conservation. Distributive issues are settled via financial incentives that try to capture the best value for money.

Australian BushTender programme (DSE, 2005a, b and c)

BushTender is an auction-based approach used in the Australian state of Victoria to improve management of native vegetation on private land. Under this system, landholders competitively tender for contracts to improve their native vegetation. Successful bids are those that offer the best value for money, with successful landholders receiving periodic payments for their management actions under agreements signed with the Department of Sustainability and Environment (DSE). These actions are based on management commitments over and above those required by current obligations and legislation. The BushTender programme ensures that priority native vegetation on private land is targeted in a cost effective manner and provides landholders with an opportunity to generate a regular and reliable income stream from their native vegetation (*www.dse.vic.gov.au/DSE/nrence.nsf/*).

The BushTender process begins with a site visit by a BushTender field officer. The field officer assesses the significance and quality of the native vegetation and discusses management options with the landholder wishing to enter into a management agreement with the regional body. Whilst on the site the field officer scores the habitat management services being offered by the landholder based on the discussed management actions. Typical activities include protection of the native vegetation (*e.g.* leaving logs lying on the ground, restricting grazing, constructing fencing) and/or active management of key habitat components (*e.g.* weed control, fire prevention, supplementary planting of native plant species). Landholders identify the actions they propose, and with the field officer prepare an agreed management plan as the basis for the bid. The Habitat Improvement Score in the BushTender programme helps to ensure that redundant elements of the scheme are reduced. In other words, it ensures that the scheme actually delivers real additional improvements in biodiversity.

The landholder is also provided with a site conservation score (how important is their site) and their habitat management services score (how much service is the landholder providing) to assist them with preparing a bid. It is then up to the landholder to determine the payment they require to undertake the proposed management actions which she/he then submits a sealed bid.

Bids are assessed on the basis of *a)* current conservation value of the site; *b)* amount of service offered: estimated improvement in vegetation condition and/or security; and *c)* cost. Funds are allocated on the basis of the "best value for money"[4] (DSE, 2005a). Successful bidders sign either a five-year management agreement only, or a five-year plus permanent protection agreement based on the previously agreed management plan (DSE, 2005b).

In 2002-2003, the BushTender scheme was tested in the Australian Gippsland region. In total 73 bids were received from 51 landholders (some landholders having bid separately on each of their sites); 33 successful landholders signed management agreements with the Department of Sustainability and Environment. The AUD 800 000 set aside for the Gippsland trial management agreements has been allocated and management agreement periods of three or six years were offered to landholders, with the further option of ten-year or permanent protection covenants. Of the successful bidders, all but one opted for at least a six-year management agreement period, with almost half of all bidders committing to further protection. In total, 1 684 ha of vegetation have been protected in the Gippsland trial, approximately half of which is considered to be of high or very high conservation significance (*www.dse.vic.gov.au/DSE/nrence.nsf/*). Follow-up visits revealed that the majority of landowners have complied with the agreements; only a small percentage have been required to complete additional work before being authorised to collect their next payment (DSE, 2005c).

This is a good example of how biodiversity goals can be achieved through a competitive process among private landowners. In this way, efficiency (best value for money) and effectiveness (habitat management) are both achieved, and distributive issues are handled through payments.

Conservation banking in the USA (summarised from Cooperative Conservation America, 2005)

Conservation banking is a new and growing tool to conserve habitats and species, and to deal with the distributive effects of conservation in an innovative way. Money is channelled from developers who would like to use biodiversity rich areas to those who protect and manage natural habitat on their properties. This way, distributive issues are solved through a market-based mechanism.

Conservation banks are lands that are permanently protected and managed, to mitigate for the loss elsewhere of listed species and habitats. Any landowner, public or private, can participate in the programme, but federal lands may require special consideration. The main advantage for the landowner is that he/she retains title to his/her land while making money selling mitigation credits. Conservation banks ensure that a given level of biodiversity will be maintained while making development possible that might otherwise compromise a species.

By June 2005, more than 45 conservation banks had been approved in the US. Banks in Alabama, Arizona, Colorado and Texas covered more than 14 000 hectares, including habitat for more than 40 species (such as the vernal pool tadpole shrimp, California red-legged frog, valley elderberry longhorn beetle and several plant species).

Nearly 20% of the population of Mobile County, Alabama lives below the poverty line. Low incomes have created a demand for small, affordable house lots, fragmenting the longleaf pine habitat essential to the threatened gopher tortoise. The Mobile Area Water and Sewage System (MAWSS) proposed that the county establish a conservation bank on properties around its reservoir. The bank is a large piece of property that can be managed more effectively

PEOPLE AND BIODIVERSITY POLICIES – ISBN 978-92-64-03431-0 – © OECD 2008

than a smaller number of unconnected tracts. Landowners and developers with tortoise habitat on their property could purchase credits to fund management of the bank and to continue building. In 2001, about 89 hectares of longleaf pine habitat became the gopher tortoise conservation bank. Today more than 55 bank credits have been sold for USD 3 500 per credit. As the bank's owner, MAWSS finances tortoise conservation on its land while helping to avoid a costly endangered species controversy. Since the bank began, the number of resident tortoises has grown from 12 to more than 60. The bank has been equally successful for developers and home buyers. The moratorium on building permits was lifted, allowing construction of affordable housing to continue.

Canada's Ecological Gifts Program (adapted from Environment Canada, 2005)

Canada's Ecological Gifts Program provides a way for owners of ecologically sensitive land to protect nature and leave a legacy for future generations. Distributive issues are settled through tax reduction. Made possible by the terms of the Income Tax Act, it offers significant tax benefits to landowners who donate land or a partial interest in land to a qualified recipient. Recipients ensure that the land's biodiversity and environmental heritage are conserved in perpetuity.

The Ecological Gifts Program is administered by Environment Canada in co-operation with dozens of partners, including other federal departments, provincial and municipal governments, and environmental non-government organisations. Thanks to this team approach and a dedication to continuously evolving and improving, the programme has become more successful each year. Since its inception in 1995, hundreds of Canadians have donated ecological gifts valued at a total of more than USD 110 million. Nearly half of these eco-gifts contain habitats of national, provincial, or regional importance, and many include rare or threatened habitats that are home to species at risk.

The programme provides a non-refundable tax credit or deduction to donors, and a reduction in the taxable capital gain realised on the disposal of the property. Corporate donors may deduct the amount of their eco-gift directly from their taxable income, while the value of an individual's eco-gift is converted to a non-refundable tax credit. The tax credit is calculated by applying a rate of 16% to the first CAD 200 of the donor's total gifts for the year and 29% to the balance. In most provinces, a reduction in federal tax will also reduce provincial tax. Unlike other charitable gifts, there is no limit to the total value of eco-gift donations eligible for the deduction or credit in a given year. Further, any unused portion of the donor's gifts may be carried forward up to five years.

Donors who dispose of capital property, such as land, may realise a capital gain (a portion of which is taxable). The capital gain arises where the deemed proceeds of disposition exceed the property's adjusted cost base (usually the original purchase price of the land). This is generally the amount by which capital property appreciates in value while it is in the owner's possession. While for most gifts the taxable portion is 50% of the capital gain, in the case of an ecological gift it is only 25%. Donors can also reduce their capital gain by lowering the designated amount of the gift to somewhere between its fair market value and its adjusted cost base. This designated amount will also be used to calculate the tax benefit.

7.3. International solutions for dealing with distributional issues

7.3.1. Global Environment Facility

The Global Environment Facility has been a major new source of funding for conservation. As shown in Table 7.3, over 3 500 biodiversity projects were funded by the GEF between 1991 and 2003. Total GEF funding was USD 1.4 billion, and over twice that amount was generated in co-financing for these biodiversity activities. GEF funding averaged over USD 100 million per year for biodiversity activities in developing countries.

Table 7.3. **GEF projects and funding, 1991-2003**

Project type	Number of projects	GEF funding (million)	Co-financing (million)
Full-sized projects	206	USD 1 438	USD 3 100
Medium-sized projects	130	USD 104.4	USD 182.3
Enabling activities	269	USD 84.8	USD 20.1
Small grants programme	3 076	USD 63.0	USD 64.6

Source: Dublin *et al.*, 2004.

Figure 7.1 shows the number of GEF biodiversity projects by implementing agency (World Bank, United Nations Environment Programme, UNEP, and United Nations Development Programme, UNDP). Except for 1994, a year of funding replenishment, there has been a general upward trend in the number of biodiversity projects approved each year. The World Bank has more biodiversity projects than either of the other two implementing agencies. These figures come from an external evaluation of the GEF biodiversity programme, which concluded that "The GEF support to protected areas has been steadfast and unprecedented. Furthermore, the GEF has also contributed to improving the enabling environments in which biodiversity conservation and sustainable use occurs" (Dublin *et al.*, 2004). While funding from GEF and other international sources has increased in recent years, the project-oriented nature of this funding means it is subject to funding cycle limitations. After

PEOPLE AND BIODIVERSITY POLICIES – ISBN 978-92-64-03431-0 – © OECD 2008

Figure 7.1. **GEF biodiversity projects approved, fiscal years 1991-2001**

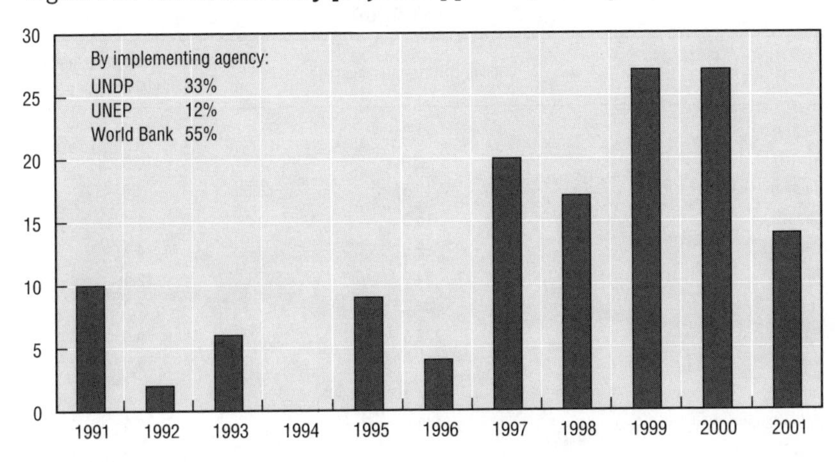

Source: Dublin, *et al.*, 2004.

the end of the project cycle, generally five to seven years, financial support for particular projects ends. Hence, the GEF financing of projects is helpful, but not sustained.

7.3.2. Transfers from overseas development assistance

According to recent estimates, the annual volume of funds flowing from multilateral and bilateral donors to developing countries for protected areas for the year 2000 was USD 370 million (Balmford and Whitten, 2003), although there has been a trend away from linking official development assistance (ODA) to the environment.

Nevertheless, significant funding is channelled to developing countries for biodiversity-related objectives. For example, between 1998 and 2000, bilateral aid for biodiversity, sustainable use, or "aid targeting the CBD objectives" provided by OECD countries averaged some USD 995 million per annum (Table 7.4).

As the table shows, the money donated is substantial when aggregated over major OECD countries. While it is difficult to gauge whether this money is adequate to cover the redistributive impacts of biodiversity policy, some anecdotal evidence suggests that it is not.

There is also little evidence that current international agreements (such as the CBD) have had discernable positive effects on halting the general degradation of biodiversity in developing countries. Balmford and Whitten (2003) point out that even at this volume, international funds leave a significant "funding gap" in international conservation between current receipts of USD 750 million and operating costs of USD 2 250 million, not

Table 7.4. **Average annual bilateral biodiversity ODA reported to the OECD, 1998-2000**

	Million USD (annual average)	Million USD % of total bilateral ODA (annual average)
Australia	21.3	2.7
Austria	2.0	0.5
Belgium	19.5	3.9
Canada	15.3	1.4
Denmark	29.8	4.5
Finland	24.9	12.1
France	44.7	1.7
Germany	275.6	9.0
Ireland	2.2	4.9
Japan	144.1	1.4
Netherlands	146.9	6.9
New Zealand	0.8	0.8
Norway	91.2	10.3
Spain	14.5	1.4
Sweden	38.3	3.9
Switzerland	15.9	2.4
United Kingdom	23.9	0.7
United States	84.2	1.0
Total	**995.1**	**2.7**

Source: OECD (2002).

counting an additional USD 5 000 in opportunity costs of current reserves in developing countries not covered by outside sources.

There is some evidence that biodiversity-related aid is declining (Emerton *et al.*, 2005; Cleary 2006). An important reason for the reduction in that aid is a move away from aid that was tied to conditions set by the donor. During recent years many countries adopted the notion that aid should be given without preconditions, so that recipient countries would be free to follow their own priorities. This change should not necessarily affect the amount spent on redistributive objectives, but would make it harder to account accurately for the amount of aid that targets CBD objectives.

While the funding gap is a fact, it cannot explain why present funding efforts seem to encourage little conservation effort, despite the presence of a global institutional regime. Considerable interest exists in the question and various explanations for this failure have been advanced. They can be divided into three broad themes:

1. Government failure and problematic distributive effects in various forms, such as perverse subsidies that compete with conservation measures and hamper their effectiveness (Margulis, 2004).

2. The persistence of dysfunctional property rights (Southgate *et al.*, 2000) and the lack of complementary rights for farmers or landowners whose land supports biologically valuable resources (Droege and Soete, 2000).
3. The insufficient pass-through of funds received under the payment mechanisms of the CBD from national governments to those individuals who actually make the day-to-day conservation decisions (Day-Rubinstein and Frisvold, 2001).

Other recurrent theories are that land speculation destroys incentives for conservation (Margulis, 2003) and that widespread corruption prevents any significant funds from reaching local decision-makers (Smith *et al.*, 2003). Domestic institutional constraints are therefore important in determining who actually benefits from international funds for biodiversity protection.

7.3.3. Private/NGO funding

Other sources of transfers that are related to distributive impacts are private and non-governmental. A number of large organisations provide substantial amounts of funding for conservation. Khare and Bray (2004) reported that during the late 1990s, private philanthropic foundations spent roughly USD 150 million globally and the private sector itself about USD 20 to USD 30 million on conservation. Chapin (2004) reported that three exceptionally large NGOs (WWF, Conservation International, and The Nature Conservancy) collectively spent USD 490 million during 2002 for conservation activities in developing countries. In some cases this involved the purchase of logging rights – which were not exercised – so the trees remained in the forest.

Private sector financing for biodiversity protection has taken three main forms: contributions to conservation trust funds, privately financed and managed reserves, and payments for ecosystem services. Foreign conservationists have participated in these schemes and have even purchased land outright for private conservation. Unfortunately, there is no guarantee that the money spent is providing the highest biodiversity benefit. It also sometimes results in problems associated with resentment of foreign land control and disputed land claims.

Debt-for-nature swaps have in the past been significant sources of transfers to some developing countries but these appear to be slowing in recent years.

Notes

1. Using a broad definition of "institutional change" as changes in implied property rights.
2. Compensation schemes in this volume include those where a landowner or user must accept restrictions on his/her land, for which they are compensated. In this scheme there is no negotiation with the landowner or user – the restrictions are imposed by regulation.
3. *www.mmm.fi/metso/international.*
4. Calculated based on the benefits biodiversity index as follows: (conservation value score X habitat improvement score)/cost required by the landholder.

ISBN 978-92-64-03431-0
People and Biodiversity Policies
Impacts, Issues and Strategies for Policy Action

Chapter 8

Combining Institutional and Procedural Approaches: Community Involvement in Management Decisions

We now turn to the most profound method for handling distributive issues: the active involvement of indigenous and local communities in the management of biodiversity. This approach combines the procedural elements of communication and participation with the institutional elements of creating rights and ownership in the implementation of the policy. Such an approach dilutes the power and influence of the policy-maker to a significant extent: participation in or even devolution of ongoing management decisions to stakeholders mean that the policy-maker sacrifices control over policy implementation. This can even result in fundamental changes to the policy itself.

As we have seen, distributive problems can arise if nature conservation management is practised with the exclusion of local communities. Thus, many distributive issues can be overcome if local communities are involved in biodiversity management. Involvement and respect of local and indigenous communities in wildlife management are generally-accepted principles that are enshrined in the Convention on Biological Diversity (CBD, 1992, article 8j). There are important distributive impacts in those principles: local communities are able to influence the nature management decisions which affect their lives.

Local and indigenous groups, however, may not be homogenous in their interests. They may have different goals and social objectives that distinguish them regionally, nationally and even internationally. For example, in developing countries poverty reduction and meeting basic needs are likely to be more important than in developed countries, where reducing local unemployment, and sharing economic and other benefits of biodiversity management with the community, might be more important (Roberts and Gautam, 2003).

The requirements for successfully involving indigenous groups and local communities differ from those of other stakeholders in important ways, including:

- **Creating a supportive legal and policy framework** which can legitimise the involvement of local and indigenous communities in biodiversity-related management. Resource ownership, access and user rights, and management plans need to be addressed, as well as the potential for community involvement and collaborative management (Gawler, 2002). A

PEOPLE AND BIODIVERSITY POLICIES – ISBN 978-92-64-03431-0 – © OECD 2008

detailed policy plan or strategy for community involvement can also guide the implementation.

- **Training policy-makers, agencies and park managers** in working with local communities. Governments often lack close relationships with the affected groups and rely on local authorities and agencies that have closer contact with local communities. But these representatives of government authority are not always well-versed in the culture, traditions and working habits of these groups. Preparation for any policy initiatives might therefore include assurances that there is some ability to interact with, and respond to, these groups.
- **Building community capacity for involvement:** local communities may lack the sophistication to work with nature conservation agencies or national park directorates. Training and education may be required for the agency personnel in management and working techniques so there is a capacity to represent and act in the interest of the community.
- **Incorporating conflict resolution mechanisms:** since communities that are going to be involved in nature management may be heterogeneous and have different interests, it is worth understanding potential sources of conflict at an early stage. Understanding cultural and social characteristics of local groups is important for choosing appropriate strategies. In managing natural resources, traditional and local communities may have some issues that deserve attention, namely those surrounding gender, power and equity within the community, traditional ecological knowledge, and the tension between short and long-term goals.
  - **Gender issues:** In many traditional communities, women and men have roles and tasks that help give society its structure. They may also have different perceptions of the need and opportunity to engage in the management of natural resources. There are many factors that influence women's capacity to engage in public work: *e.g.* household status, employment, work related rights, double work burden, education and literacy, health, ability to control fertility, access to financial resources, existence of legal rights, traditions and cultural values, socialisation and self confidence (Buchy *et al.*, 2000). Participation of women in decision-making is usually a sensitive issue in traditional communities, and sufficient time is needed to overcome cultural barriers. However it has generally been associated with positive outcomes because women are more likely to rely on nature for their day-to-day activities.
  - **Power and equity within the community:** communities interacting strongly with nature often have hierarchical structures which mean that some people or families have better access to resources than others, *e.g.* chiefs, wealthier families, families with private property or animals

and households closer to the natural areas. These groups may thus enjoy greater benefits from the existence of a natural area and their positions allow them to act more strongly for their private interests. Therefore when involving local people in wildlife management, it is important to ensure that the poorer and more vulnerable groups have a voice in the decision-making bodies.

- **Traditional ecological knowledge:** local and indigenous communities may have specialised ecological knowledge and traditions that can be useful for biodiversity management. Knowledge of natural processes, species identification, seasonal productivity of certain species and influencing circumstances are often embedded in local culture. Traditional practices, *e.g.* voluntary restrictions on access and use of certain areas, sacred or no use sites, zoning, taboos in certain seasons, minimum size of stock to be saved, etc., may have evolved through long traditions of experimentation and experience (Berkes, 1999; Gawler, 2002). This valuable knowledge can be combined with modern scientific methods and form the basis for joint work and monitoring.
- **Short and long-term goals:** an important caveat to community or local management can be the problem of maintaining behaviour that is focused on long-term outcomes. For example, assigning private rights over a resource that had been public and where social norms had made that public resource sustainable, may induce shorter term behaviour centred on private interests. "Selling-out" to commercial interests may be in an individual's interest, but not in the community's – the rate at which cash-poor individuals discount the future will be higher than that of societies. This may lead to destructive land use (*e.g.* burning of forests to clear land, or selling it to private commercial interests, Chapin, 2004) that neglects the public aspect of the resource.

8.1. Forms of community involvement

There are many forms and degrees of involvement by communities. They are influenced by traditions within communities, systems of property rights, and even by the administrative authority within a country. In the following section we distinguish three main forms: community-based management, joint management of natural resources by communities and government agencies, and management by stakeholder bodies.

8.1.1. *Community-based management*

One response to the need for community involvement has been the implementation of community-based management or collective management in which land, or a biodiversity-rich resource, is a common resource managed

 PEOPLE AND BIODIVERSITY POLICIES – ISBN 978-92-64-03431-0 – © OECD 2008

by the local community. For this to work well, good ties are needed between and within communities.

Community-based management is more common in developing countries, where a greater number of traditional settlements exist. Some examples can also be found in those OECD countries where Aboriginal or indigenous communities still live in reserves or in protected areas (*e.g.* Canada, USA, Mexico, Finland and Sweden). The rights of these people and their participation in management are key in many of these countries.

There are several ways that community-based management can be organised and implemented. Local administrative and management bodies (*e.g.* the village council) might be set up by local people to prepare and implement the management plan for the area. Access and user rights (*e.g.* fishing, hunting, collection of wood) can be created by the government where the administrative body is empowered with the assignment of these rights. In some cases, subsidies are also provided by the government to compensate for lost opportunities. Revenues from the area (*e.g.* tourism, trophy hunting) can be used to support conservation objectives.

Empirical evidence on equity and distributional benefits is rather mixed when it comes to respecting the rights of indigenous and local groups to manage resources themselves. It is important to specifically account for those who are included and excluded from the decision-making body. It is also important to delve into traditions of resource use and, where necessary, even put in place restrictions if economic incentives favour destructive use. It might be useful to explore other forms of land use such as private land for individuals from the community. The inclusion of poorer and/or vulnerable resource users (*e.g.* women and youth) in community management as well as decision-making bodies also has been shown to be important for equitable benefit sharing (Mahatny and Russel, 2002; Adhikari *et al.*, 2004).

8.1.2. *Joint management of natural resources by community and governmental agency/park administration*

In joint management of natural resources, local communities and administrative bodies share some management responsibilities. It is important that the tenure, ownership and user rights over the resource are clear. This form of management is most suitable when the area is under the direct control of the parties (Buchy *et al.*, 2000). The rights and responsibilities can be laid down in a contract between the conservation authority/body and the local communities. The time frame might be very long, *e.g.* 99 years for some Australian national parks.

Joint management committees can be set up which are responsible for drawing up management plans and making decisions about park management (Reid *et al.*, 2004). They work best when there is rough equality of power and influence between the parties. Local or indigenous communities can take on certain management tasks, *e.g.* fire management, game management, monitoring of habitats. In return for these activities they are assigned certain rights, *e.g.* hunting, fishing, collecting wild plants or wood for subsistence use. In this way local knowledge and competence in nature management can be made best use of. It is important to acknowledge the difference between the working cultures of indigenous people and park managers when assigning tasks. The approach is likely to work best when local people are responsible for those tasks which are part of their culture.

Some examples of joint management are parks in South Africa and Australia (Reid *et al.*, 2004). There are also examples from North America, where more than two partners (*e.g.* the management body and local communities, plus recreational wildlife users and subsistence users) are involved in the management of the area (Buchy *et al.*, 2000).

8.1.3. *Management by stakeholder bodies*

Management by stakeholder bodies is another common way of involving local communities in the management of natural resources. In this case a set of stakeholders, including the representatives of governmental bodies, local businesses, local communities and civil organisations, form the advisory board of a natural area. They are usually responsible for making or revising strategic plans and for supervising the management of the area. The main characteristic of this form of management is that there is usually mixed ownership and no full control by any individual board member over the use of the resources (Buchy *et al.*, 2000).

This is a less intensive public participation method for biodiversity management than the other methods, but it also can be an effective way to provide benefits for the local community through assigning access and user rights, lowering entrance and user fees, selling local products and services from the area, employment possibilities or increased income. The interests and needs of local people can be expressed on the stakeholder board and through collaborative actions. This approach has been working in some countries, *e.g.* regional parks in France, watershed/catchment management in the USA.

Table 8.1 summarises the main characteristics of the three forms of community involvement; many of the examples are discussed in further detail below.

Table 8.1. **Main characteristics of the three forms of community involvement**

Characteristics	Community-based management	Joint management of community and governmental agency	Management by stakeholder bodies
Ownership of the area	Community ownership or state ownership but handing back the property or user rights to the communities	Community ownership (sometimes the land is leased back to the state) or state ownership with special community rights	Mixed ownership
Legal regulation required	For property rights, framework for community management	For property rights and need to sign a contract between the parties (may be a requirement of the contract as well)	Potential for stakeholder bodies
Degree of community involvement	The whole community participates	Large part of the community participates (both directly and indirectly)	Only part of the community participates (through representatives or with direct involvement in some activities)
Where is the balance of power crucial?	Within the community (poor, vulnerable groups, young people, women)	In the community as a whole, and within the community (poor, vulnerable groups, young people, women)	Between the stakeholders
Managing distributive issues	Fair and balanced representation is required in the decision-making body. Sometimes outside help is needed to overcome cultural barriers	Special rights need to be assigned to the community Fair representation is needed in the decision-making body	Fair and balanced representation is needed in the decision-making bodies
Examples	Saami villages, Sweden. Community-based participation in wetland conservation, West Kalimantan, Indonesia. CAMPFIRE Program of Zimbabwe	Co-management schemes in Aboriginal national parks (*e.g.* Kakadu, Australia)	Waswanipi Cree Model Forest, Canada Community forest partnership, England Watershed management with community participation (Conasauga River Watershed), USA Regional nature parks, France Wetland co-management in the Djoudj National Park, Senegal

8.2. Facilitating community involvement

There are many ways the government or its bodies can foster or facilitate community involvement in the management of natural resources. Some examples are as follows:

- **Technical assistance:** local communities may lack training in assessment, management or monitoring. Assistance can be provided with scientific knowledge, models, techniques (*e.g.* geographical information systems) or the use of modern equipment. Guidelines can be prepared and training can be organised to help the communities.

- **Co-ordination:** community involvement can be more effective if it is part of a nationally-organised framework: *e.g.* community forest programmes, watershed or catchment programmes or Aboriginal management programmes. Some examples include Canada's Model Forest Program, England's National Community Forest Partnership and the USA Watershed Protection and Restoration Program. When such national frameworks exist, experiences at the local level can be shared more widely, *e.g.* through regional and national discussion forums.
- **Financial assistance:** community-based, shared or stakeholder management can be aided by financial assistance. Large restoration projects especially might need financial support to be successful. Projects sometimes need seed money to start a co-operative operation (*e.g.* paying the members of the decision-making body). In some countries (*e.g.* the US or Canada) grant programmes are launched to help these community-based efforts. In other countries, benefits from the area partly go to local communities: *e.g.* park fees in Uganda, buffer zone fees in Nepal and tourist revenues in the CAMPFIRE programme, Zimbabwe.
- **Clearinghouse mechanism:** a clearinghouse mechanism can help spread information about local and regional experiences, or the results of projects or discussion forums.

8.3. Examples of different forms of community involvement

Below are just a few examples of the many different types of management and state assistance both in developed and developing countries.

8.3.1. Community-based management examples

Rights of Saami people in the World Heritage site, Lapponia, Sweden (summarised from Lusty, 2000)

The Lapponian Area covers almost 9 400 km^2 and lies in Norrbotten county, in the circumpolar zone of Northern Sweden. It is inhabited by the Saami people, who arrived in the area between 4 and 5 000 years ago. For thousands of years, the Saami lived mainly by hunting wild reindeer for fur and food. They led nomadic lifestyles, following the reindeers' annual grazing cycles. A few Saami families still migrate and maintain their summer residence in small cabins. The majority, however, now lives in villages. They have a rich folk culture with traditional handicrafts, clothing and music, which, together with their language, are distinctively different from those of other ethnic groups in Scandinavia. The Saami people's rights are protected by laws dating back to 1886. All reindeer breeders belong to a Saami village, which represents an administrative and economic unit. The members decide

PEOPLE AND BIODIVERSITY POLICIES – ISBN 978-92-64-03431-0 – © OECD 2008

how herds are to be managed within the confines of the Reindeer Husbandry Act (see Section 6.4.2), which sets a maximum allowance of 280 000 reindeer for the whole of Sweden. The Saami village can also decide how many reindeer each of their individual members is allowed to keep. There are government subsidies available for herdsmen, based on kilograms of meat. Saami also have fishing and hunting rights.

This is a good example of how an indigenous community can have rights to use and manage natural resources within the rules of the state (*e.g.* maximum allowances) and with the state's financial support (subsidies to herdsmen). Distributive issues are settled between the state and the community and also within the community (see also the conflict case in Section 6.4.2, which describes how these rights were violated).

Customary rules in community-based wetland conservation, West Kalimantan, Indonesia (Wickham, 1997)

The Danau Sentarium Wildlife Reserve comprises 125 000 hectares of lakes and temporarily and permanently flooded lowland forest in the north-central region of West Kalimantan, Indonesia. Water levels fluctuate during the year, and there are three months without any water at all. The reserve supports a diverse flora and fauna, and unique habitats. Around 3 500 people live in 40 permanent and seasonal villages within the watershed. Research in the area showed that customary rules and regulations for resource use and sanctions for breaking them have existed in the communities for centuries (Wickham, 1997). Those that are in line with current regulations could be an integral part of community-based nature conservation strategies relying on self-regulation. Around 40 such rules were identified in the research, some of which are listed in Table 8.2.

Table 8.2. **Overview of various regulated resources in Danau Wildlife Reserve**

Forest resources regulations	Fishing equipment regulations	Selected fish regulations
Honey	Fish nets (type/size)	Jelawat (*Leptobarus hoeveni*)
Rattan	Fish traps (type/size)	Betutuk (*Oxyeleotris marmorata*)
Hunting	Other fish equipment	Siluk (*Scleropages formosus*)
Forest fires	Fishing with electricity	Toman (*Ophicephalus micropeltes*)
Logging	Fishing with poison	

Source: Wickham, 1997.

This case is a good example of where traditional restrictions on the use of nature in a community can be used to set rules for community-based nature management. If these restrictions and self regulation are accepted by the community, no distributive problems are likely to arise.

Zimbabwe's CAMPFIRE Programme (Alexander and McGregor, 2000; Jones and Murphree, 2001; Mashinya, 2007)

Early conservation laws in Zimbabwe outlawed hunting and prohibited local communities from managing or benefiting from wildlife. Private farm owners were given the right as "appropriate authorities" to use wildlife on their land by the Parks and Wildlife Act of 1975, while users of communal lands* were not. This led to conflicts between the government who "owned" the wildlife on communal land, and the people residing on that land who were not allowed to use the wildlife for their subsistence, and who also suffered damage to their crops or livestock by wildlife. The Park and Wildlife Act was amended in 1982 to allow "appropriate authority" status to be granted to local rural district councils (RDCs), enabling them to legally exploit natural resources within their jurisdictions.

The CAMPFIRE programme (Communal Areas Management Programme for Indigenous Resources) was developed after this amendment to promote greater local control over the management and use of biological resources in communal areas. This programme sought the participation of local communities in generating wildlife revenues through sustainable use, rather than simply being the passive recipients of money via RDCs (Alexander and McGregor, 2000). Due to the previously rapid conversion of wildlife habitat to agriculture and grazing, there was interest in creating economic incentives for preserving wildlife and its habitat. The programme had several objectives, including voluntary participation by communities in developing long-term solutions to resource management problems; introducing new systems of group ownership and rights to natural resources for resident communities; providing appropriate institutions for resource management and exploitation by resident communities for their direct benefit; and providing assistance to communities wishing to join the programme. The project was also designed to provide money from tourists and both meat and revenue from trophy-hunters (Young *et al.*, 2001). At least 50% of these revenues were to go directly to communities (Jones and Murphree, 2001).

Despite the appealing goals of this programme, its implementation has been criticised (Alexander and McGregor, 2000). Recent research shows that after donor funding ended in 2000 and Zimbabwe's severe national political and economic crises began, the extent and quality of community participation has declined sharply and benefits were captured by local elites. The loss of NGO support has also had negative effects on the success of the programme (Mashinya, 2007).

* Areas which were held in trust by the government for indigenous tribes on a collective basis.

PEOPLE AND BIODIVERSITY POLICIES – ISBN 978-92-64-03431-0 – © OECD 2008

The CAMPFIRE programme was a brave attempt to revitalise community-based biodiversity management in a way that also addressed distributive issues (*e.g.* creating use and benefit-sharing rights). However, democratic instability and the withdrawal of international financial support can have important negative effects on both process and outcome.

8.3.2. Joint management between community and/governmental agency: some examples

Contracts with Aboriginal people in Kakadu National Park, Australia (Grady, 2000; Reid et al., 2004)

Kakadu National Park (Table 8.3) is situated in the northern part of Australia and covers 19 804 km^2. It is also a World Heritage Site. Approximately 50% of the land in the park is held as inalienable freehold land by Aboriginal groups. The Aboriginal people have been continuously present in the area for more than 50 000 years. Having lost their lands to newcomers, they were reinstated in a 1976 act of government. The estimated number of Aboriginal people in the area was 1 200 in 1991. There are about 16 clans of traditional owners widely scattered throughout the park. New legislation, the *Environment Protection and Biodiversity Conservation Act (1999)*, recognises the critical role of indigenous people in the conservation and sustainable use of ecological resources, and in holding traditional knowledge.

Since the act came into force, contracts have been signed with the Aboriginal groups governing management of the area. Parks Australia (the governmental agency managing national parks) and the Aboriginal traditional owners jointly manage the park, and Parks Australia covers the cost of it. The role of the Aboriginal groups in the management and administration of the

Table 8.3. **Key characteristics of Kakadu National Park**

Characteristics	Values in Kakadu National Park
Contract signed and duration	Stage I. 1979, Stage II:1991, Stage III.1987,1989,1991 (100 years)
Size	1.9 million ha
Vegetation	Rainforest, grasslands, wooded savannas, eucalyptus forests and mangroves
Owners	Bininj/Mungguy traditional owners (about 200-300 people represented by three Aboriginal land trusts)
Conservation authority financial benefits and costs	Costs AUD 11 million to AUD 14 million per annum to manage and government provides 74% of the park budget
Financial benefits to landowners	Lease money and 39% of income from tourism (totalling AUD 1.3 million in 2000)

Source: Reid *et al.*, 2004.

park is laid down in the management plan. Their former advisory role has become a more formal management role. Five local Aboriginal associations are set up in Kakadu, representing the different political interests of different clans, and they oversee aspects of financial investment, local business, enterprise ventures and other businesses for their members. The Aboriginal people are involved in the management of fire, the native vegetation structure and habitats. Their traditional knowledge of land management is critical for sustaining the habitats. They are also able to practise their traditional rights of gathering native plants for food and handicrafts, and of hunting and fishing. They consult with governmental bodies about the sustainable take levels of different species (Grady, 2000; Reid *et al.*, 2004).

The operation of Kakadu National Park is a good example of co-management and benefit-sharing with the Aboriginal community. Distributive issues are settled in the contract (participatory management, rights to use the area and sharing the revenues).

8.3.3. Stakeholder management examples

Canada's Model Forest Program (Canadian Model Forest Network, 2006)

Canada's Model Forest Program was launched in 1992 by the Government of Canada through the Canadian Forest Service (CFS). The programme is one of the world's largest experiments in sustainable forest management. A model forest is an area where the latest forestry techniques are researched, developed, applied and monitored. It operates through a grassroots partnership that includes a variety of stakeholders who value the forest for different reasons. Canada's Model Forest Program currently involves 11 model forests ranging in size from just over 100 000 hectares to nearly 8 million hectares.

The main objectives of the programme are: *i)* to increase the development and adoption of sustainable forest management systems and tools within and beyond model forest boundaries; *ii)* to share knowledge gained through the programme at local, regional and national levels; *iii)* to strengthen model forest network activities in support of Canada's sustainable forest management priorities; and *iv)* to increase opportunities for local-level participation in sustainable forest management.

Model forests build partnerships with a wide range of individuals and organisations whose interests in the forest may vary, but who share the common goal of sustainable forest management. Partners include: scientists, Aboriginal communities, environmentalists, forest industry, community groups, landowners, national parks, academic institutions, governments, recreation enthusiasts and others interested in sustainable forest management. Partners invest significant time, effort and resources learning

PEOPLE AND BIODIVERSITY POLICIES – ISBN 978-92-64-03431-0 – © OECD 2008

about and appreciating each other's views and expertise. This allows consensus-driven partnerships where decision-making is shared to achieve social, environmental and economic sustainability in forest management.

Each model forest is managed by a partnership made up of local individuals and organisations. Their goal is to make sure that the forest continues to be a healthy and dynamic part of their community. Successes at the local level can then be shared with other model forests through Canada's Model Forest Network. The success of Canada's Model Forest Program has attracted worldwide attention. An International Model Forest is now in place, with 20 model forests in 15 countries. Several other countries have also expressed an interest in joining the network.

Canada's Model Forest Program is a good example of stakeholder management. The participatory decision-making method addresses distributive issues and helps find the best solutions for all stakeholders. Networking and information sharing are also useful elements of the programme.

Waswanipi Cree Model Forest, Québec, Canada (Roberts and Gautam, 2003; Pelletier, 2002)

In Canada there is a legal basis allowing local communities to sign forest management agreements with provincial governments to create a community forest. The Waswanipi are local tribes in Québec who successfully operate a community forest management system called the Waswanipi Cree Model Forest. Their vision is to link traditional tribal ties with the development of resource-based activities, such as forestry, tourism and recreation. It tries to combine traditional ecological knowledge with applied research and technologies to develop new sustainable forest management practices (Roberts and Gautam, 2003).

Located 800 kilometres north of Montreal, Waswanipi is the southern-most of the Cree communities in Québec. The people of Waswanipi have lived in the boreal forests since time immemorial. Their land base extends over 35 000 square kilometres and is divided into 52 ancestral family hunting territories, called trap-lines. The Crees have benefited from the boreal forest for millennia, while successfully maintaining a healthy and viable economy based primarily on hunting, fishing and trapping. It is only recently that outsiders have seen the potential for extracting natural resources and forestry companies have established a permanent presence in the area.

The Waswanipi Cree Model Forest is a special project where community participation, sustainable forest management and community/technology transfer play a major role. A Working Committee (of 20 people from 13 different organisations) was created to make strategic decisions for the project. Crees favoured co-management, where they participate at all levels of forest management planning (laws and regulations, 25-year plan, 5-year plan, yearly plan) and monitoring. The co-management approach was accepted by the committee as a means of improving forest management planning. After setting the main tasks, a Development Team was created which involved representatives of three forestry companies, the Government of Québec, and local communities. The learning experience has been successful, and many problems (*e.g.* communication, balance of power, timing) have been gradually overcome or mitigated (Pelletier, 2002).

Although the Waswanipi operate a community forest, this case shows that the operation can be improved by involving more stakeholders. Through negotiation and the participatory planning process, distributive issues have been raised and settled because the plans have been accepted by all the groups involved.

England's National Community Forest Partnership (www.communityforest.org.uk)

The National Community Forest Partnership is made up of 12 Community Forests in England with 58 local authority partners, the Forestry Commission and the Countryside Agency. The 12 forests are located in and around major towns and cities, with each forest working with the local authorities, government agencies and a variety of partners within their operating area. The Community Forests all benefit from a dedicated local team or organisation working with a variety of partnerships and delivery agencies to carry out projects in the area. They are particularly effective in the protection and management of sensitive areas like semi-natural woodland, moss-lands, heather moorland and wildflower areas, river systems, unimproved grassland, Sites of Special Scientific Interest, Sites of Biological Importance and Local Nature Reserves. Involvement of local people in planning and implementation and their training is an important part of the programme.

The community forests are good examples of stakeholder management, where local people participate as well. Local communities benefit from the improved state of local forests, and they probably voluntarily contribute to the costs of the projects.

Watershed management with community participation in the USA (EPA, 2001)

The Clean Water Action Plan was announced in the USA in 1998 to improve water quality nationwide. The action plan seeks to support existing local watershed partnerships to address critical local problems, develop

PEOPLE AND BIODIVERSITY POLICIES – ISBN 978-92-64-03431-0 – © OECD 2008

restoration strategies and implement solutions that improve the watersheds' health. A watershed (also known as a catchment or basin) is a geographical area in which all the falling water drains to a common water body, *i.e.* river, lake or stream. The watershed approach uses watersheds to co-ordinate the management of water resources. It integrates biology, chemistry, economics and social considerations into decision-making. A successful watershed approach includes the support, participation and leadership of local stakeholders and land users. A watershed approach recognises needs for water supply, water quality, flood control, navigation, hydropower generation, fisheries, biodiversity, habitat preservation and recreation, and recognises that these needs often compete. It addresses natural resource issues that cross jurisdictions and political boundaries (EPA, 2001; Clean Water Action Plan, 2000).

Seven themes of watershed management are commonly found: *a)* increasing public education and awareness; *b)* developing new partnerships and co-ordinating efforts; *c)* collecting necessary information through monitoring and research; *d)* establishing appropriate plans and priorities; *e)* obtaining funding and technical assistance; *f)* implementing solutions; and *g)* evaluating the results (EPA, 2001).

There are over 3 000 local watershed groups. Watershed partnership can include any person or group interested in watershed health, *e.g.* landowners, elected officials, representatives of federal, tribal, state and local government agencies, agricultural organisations, business organisations, environmental organisations, student groups and senior citizen organisations. It ensures that activities carried out are based on mutual understanding and consensus. Various federal agencies also encourage local watershed efforts with financial and technical support. A Regional Watershed Coordination Team was established by regional offices of federal government agencies in 12 river basins. It also helps the watershed groups by co-ordinating governmental efforts (EPA, 2001).

Wetland co-management in the Djoudj National Park, Senegal (Diouf, 2002 in Gawler, 2002)

The Djoudj National Park was created in the delta ecosystem of the Senegal River in 1971. The population of the area is characterised by dispersed settlements, and there are now eight villages around the park. The main socio-economic activities are raising livestock, agriculture, fishing, handicrafts, trading and hunting. The population was removed from the area when the park was initially established, but this exclusionary policy was changed after 1994 with the introduction of a new participatory management policy. The new policy aimed to give value to defined spaces, regenerate natural resources and restore the environment, define customary law, and give value to local environmental knowledge. A five-year integrated management plan was developed through consultation with the relevant stakeholders (local populations, state technical services, NGOs, research institutes and international partners).

Four committees are responsible for the implementation of the management plan: Orientation, Scientific, Park Management and Village Conservator. The park's Orientation Committee was responsible for gathering support for the management plan, and for making the major decisions affecting the park: *e.g.* investments within the buffer zones. The Scientific Committee prioritises and approves scientific and technical research in the area and investments to be carried out within and around the area. The members of the Park Management Committee are the main stakeholders of the area, including two representatives of each village in the buffer zone. This committee influences the implementation of the management plan. Effective community involvement is secured by the operation of the Inter-Village Conservation Committee, which co-ordinates specialised committees on ecotourism, waterways, health and forestry/pastoralism. These consultation structures have facilitated a closer relationship between the local people and the park agents.

Change in the planning and operation of the Djoudj National Park illustrates how previously excluded local communities can be involved in the park's strategic planning and operation once again. Participation in all dimensions of decision-making can ensure that distributive issues are discussed and solved.

Residents' task force for water quality improvement in Korea (OECD, 2006)

The Daepho River is a 9 km-long stream flowing into the Nakdong River in Korea. Until the early 1970s, the Daepho could still be used as a source for potable water without treatment. But water quality deteriorated due to waste water discharge from nearby residential areas and local industrial firms, livestock enterprises and restaurants. In 1997, the local authority drew up a water management plan and announced its intention to designate the area as a water source protection area. Local residents protested against the restrictions, and after some negotiation an agreement emerged that if local residents could revive the river, the government might reconsider the designation.

As a result, the residents formed a "task force for water quality improvement" and started to voluntarily clean up the river. Each household contributed a certain amount of money every month to raise funds. Women's associations organised campaigns in each village to save water and reduce the

PEOPLE AND BIODIVERSITY POLICIES – ISBN 978-92-64-03431-0 – © OECD 2008

use of detergents. The city council installed settling tanks for every household and restaurant to prevent food waste discharge into the river. Livestock enterprises installed pre-treatment facilities. The task force also mechanically cleaned up the river. Artificial wetlands were planted with parsley dropwort to filter domestic waste water.

Within a year, these efforts improved the water quality of the Daepho to Class I. The previously cloudy water turned clear, enabling crayfish, endangered shellfish and other fish to return. The task force continued its efforts and in 2002 a voluntary agreement was signed in which the citizens made a commitment to maintain the water quality level and in return the government deferred the designation of the water source protection area.

The Korean case shows that voluntary joint action by citizens can be more effective than implementing a strict regulation. The final result is good water quality with increasing river biodiversity and good co-operation among citizens.

8.3.4. Benefit sharing with communities involved in nature conservation: some examples

In some developing countries, policy involves creating a protected area with restricted access and charging fees to visitors and other users for accessing the area's resources. The institutional innovation in these programmes is to channel parts of these revenues back to local communities as compensation. From a distributive perspective, the relationship of this compensation to the burden imposed on the local communities determines whether equity issues are adequately addressed. Additionally, these schemes are not unproblematic, since rather than receiving predictable streams of compensation, local communities receive flows that vary with the total revenues generated. If communities are risk-averse, the additional well-being generated by these funds will therefore be lower than their cash value.

Park fees channelled back to local communities in Uganda (Musinguzi, 2006)

The Mgahinga Gorilla National Park is home to a large variety of wildlife, including about half the world's critically endangered mountain gorillas. The government of Uganda passed a law in 1996 requiring the park authority to contribute 20% of the proceeds from park entrance fees to local communities adjacent to the park. The government did this in an effort to help local people appreciate the benefits stemming from the park and from gorilla tourism. In addition, communities near the park have had conservation training from some non-governmental organisations such as CARE. Grants have also been given for building primary schools, health clinics and improving roads.

Studies show that people's attitudes have generally improved since these initiatives were implemented.

Sharing buffer zone fees with local communities in Langtang National Park, Nepal (adapted from CBD, 2005: www.biodiv.org/doc/world/np/np-nr-me-en.doc)

In 1993, Nepal introduced an innovative management system by establishing buffer zones in and around protected areas and sharing revenue earned by national parks with local inhabitants. This was made possible by a provision made in the fourth amendment to the *National Parks and Wildlife Conservation (NPWC) Act (1973)*. According to the provision, the buffer zone communities are entitled to receive 30 to 50% of the total annual revenue generated from the protected areas.

Langtang National Park (LNP) is a good example of conservation and sustainable use of mountain biodiversity. The park, which covers 1 710 km^2, was declared in 1976 to conserve endangered species such as the musk deer (*Moschus chryogaster*), red panda (*Ailurus fulgens*), snow leopard (*Uncia uncia*) and their habitats (including the watersheds of Trishuli River and mountain pastures), as well as local cultural heritage. The other objective was to promote sustainable mountain tourism to benefit local people and improve their living conditions. The national park is located about 40 km north of Kathmandu, the capital of Nepal, and spread over three mountain districts.

The park's buffer zone was defined in 1998 and covers an area of 420 km^2, runs through three districts and includes 34 village development committees (VDCs). The government has been ploughing back 50% of the total revenue earned by the park into the buffer zone for community development activities. As of October 2005, the Buffer Zone Management Committee (BZMC) had mobilised NPR 14.1 million (1USD = NPR 71) for biodiversity conservation and socio-economic development programmes for buffer zone communities. Apart from government support, the legal provision also encourages conservation partners to complement the park's efforts. A number of national and international NGOs have also joined hands with the national park and buffer zone management council for community development activities.

This case is a good example of how distributive issues can be settled through a benefit-sharing programme. It helps raise the living standards of local communities whilst also making them more committed to biodiversity programmes.

PEOPLE AND BIODIVERSITY POLICIES – ISBN 978-92-64-03431-0 – © OECD 2008

ISBN 978-92-64-03431-0
People and Biodiversity Policies
Impacts, Issues and Strategies for Policy Action

Chapter 9

Summary and Conclusions

Successful biodiversity policies improve welfare overall by correcting the fundamental externalities of managing biologically diverse habitats and ecosystems. Within the overall improvements, however, biodiversity policies can create winners and losers. OECD policy guidelines call explicitly for a consideration of these distributive effects on the absolute and relative well-being of different groups of people. This book has presented an analysis of the distributive impacts of biodiversity policies across different groups, across different spatial scales and across time. We have offered methods for measuring the impacts and explained the relationship between policy objectives, instrument choice and distributive outcomes. We have also considered arguments from the economic literature for addressing distributive issues within biodiversity policy choice, and offered different methods for integrating distributional concerns into policy-making and for managing conflicts induced by biodiversity policies. Finally, we have presented a wealth of case studies to document both the complex chains leading to distributive outcomes, and best practice in merging efficiency and equity considerations in policy design, implementation and ongoing management.

Our main conclusions are as follows.

I. Paying attention to distributional impacts matters

- Paying attention to distributive outcomes in biodiversity policies will often maximise efficiency by permitting the policy to succeed. Policies built on excessively narrow definitions of efficiency can often lead to wasteful conflict and be ultimately self-defeating.
- There are a number of fundamental and practical reasons why biodiversity policies should include redistributive objectives. This goes against a key doctrine of welfare economics, which states that gains should be redistributed using separate policies after the biodiversity policy has been implemented. However, such separation is not always possible for biodiversity policies. One reason is the economics of market failure (*i.e.* the presence of public goods); another is the absence of property rights that *ex ante* give claims to those who are likely to be affected by policy.
- Pursuing biodiversity policies without considering their distributive consequences may involve serious net efficiency losses. This is because a policy that creates conflict may not only forego the potential gains from

the policy itself, but may also then cause other policies to start off in a confrontational mode that reduces the possibility of successful negotiation.

II. There are many ways to measure the impacts of biodiversity on welfare

- For all areas of social policy, the decision to implement a policy should be determined by the balance of benefits and costs. But when we are concerned about well-being, benefits and costs cannot be limited to monetary terms, but must include *any* impact that results from policy implementation.
- Impacts include the direct and indirect effects of both the concrete and abstract aspects of biodiversity. Those impacts need to be methodically accounted for across many economic, social, spatial and temporal groupings.
- The method chosen depends on the policy measure, the geographical scale, and on the data availability. Each of the methods has particular strengths for capturing distributive effects and weaknesses in capturing important dimensions or enabling different levels of data aggregation.
- Methods to help the policy-maker identify the main groups affected by the policy and the important distributive effects in monetary and social terms can be grouped into: a) income-equivalent measures (summary measures of equality such as the Lorenz curve, extended versions of CBA, social accounting matrix, distributive weights and Atkinson inequality index); b) alternative measures (employment or child health-based analysis); and c) multidimensional measures (stochastic dominance analysis, multi-criteria analysis and social impact assessment). The latter two groups combine quantitative and qualitative data to capture some of the complexity of distributive impacts beyond their economic dimension.
- The different methods have different data requirements. Therefore, while it may be desirable in an exhaustive analysis of distributive effects to use several measures, extending the number of measures and dimensions assessed requires additional time and resources.

III. Biodiversity policies have both primary and secondary distributional effects

- The impacts of biodiversity policies can be divided into primary (the direct impact of the policy) and secondary effects (the indirect impacts of the instruments chosen to implement the policy).
- As a rule, the greater the change brought about by the policy, the greater the primary effects. Primary effects usually imply net costs to the less well-off segments of a population. These primary impacts, however, do not represent the ultimate distributive outcomes. This is because

biodiversity policies need to be implemented and in the process of implementation, instruments need to be chosen.

- Secondary effects occur as a result of the policy instrument chosen to implement the policy. The more coercive and less reward-based the instrument, the more accentuated the secondary effects of the policy. Historically there has been considerable use of instruments which put a significant amount of the burden of conservation policies on poorer households.
- The trend towards market-based instruments in biodiversity policies is likely to reduce the progressive effects generally associated with traditional instruments. However, there is evidence that while market-based instruments do not hurt lower income households, higher income households tend to profit relatively more.
- There is a spatial mismatch between costs and benefits of biodiversity policies because benefits tend to be diffuse, while costs are locally concentrated.
- Protecting biodiversity today can also have uneven impacts over time and affect future generations differently. These problems of intergenerational equity can be addressed through hyperbolic discounting of costs and benefits arising at different points in time. At the same time, consistency between inter- and intragenerational equity is required.
- At the international level, there are still difficulties in translating developed country populations' willingness to pay for biodiversity conservation into sustainable funds to areas of high conservation importance (usually in developing countries). An additional factor is that the internationally-agreed rules for sharing global gains from biodiversity conservation do not distribute these gains fairly.

IV. Policies and instruments can reduce the distributive effects of biodiversity

Instrument choice is an important modifier of the primary impacts of biodiversity policies because it can channel gains and losses in particular directions.

A wide variety of instruments and approaches is available for mitigating and potentially reversing distributive effects. These can be divided into four categories:

- *Methodological*: use the measures listed in point II above to compute the potential aggregate welfare improvement of policies and choose instruments. This means that the policy-making process is now augmented by a consideration of distributive impacts. At the same time, the policy-maker still retains full control over information gathering, policy evaluation and choice, and instrument choice.

PEOPLE AND BIODIVERSITY POLICIES – ISBN 978-92-64-03431-0 – © OECD 2008

- *Procedural*: enrich the policy-making process by involving those individuals who will be directly affected by biodiversity policies. While diluting the policy-maker's influence, this approach allows for buy-in and ownership by affected individuals, groups, and households, thus reducing the likelihood of conflicts during policy implementation.
- *Institutional*: accompany biodiversity policies with explicit changes to the institutional structure under which individuals and groups take decisions that affect the target habitats and ecosystems. These may include creating property rights and entitlements as well as novel markets and contract schemes in order to manage distributive impacts. The institutional changes can be either voluntary, involuntary, or a mixture of the two.
- *Combined procedural and institutional approaches:* to bring about institutional changes to allow affected individuals, households and groups to become involved in policy decision-making on an ongoing or even permanent basis. This is the most profound way of addressing distributive issues as it allows various players to actively shape the design and implementation of biodiversity measures. Different forms of involvement are possible (community-based management, joint management and broader stakeholder involvement). They can be tailored to the specific circumstances of the policy context and to achieve the desired trade-off between involvement of stakeholders and control by policy-makers.

These different integration strategies are mutually compatible, but pose challenges, require resources, and need to have political support.

A key message is that there is a general shift away from recommending "one-size fits all" solutions. There is a wide and growing base of documented policy experience available in merging efficiency and equity objectives and best-practice examples for a wide variety of institutional and ecological settings. The knowledge base for policy-makers, and hence the foundation for well-informed policies in the future, is continuously expanding.

ISBN 978-92-64-03431-0
People and Biodiversity Policies
Impacts, Issues and Strategies for Policy Action

References

Adger, W.N and C. Luttrell (2000), "Property Rights and the Utilisation of Wetlands", *Ecological Economics,* 35 (2000) 75-89.

Adger, W.N. *et al.* (1997), "Property Rights and the Social Incidence of Mangrove Conversion in Vietnam", *CSERGE Working Paper* GEC 97-21.

Adhikari, B. (2002), "Household Characteristics and Common Property Forest Use: Complementarities and Contradictions", *Journal of Forestry and Livelihoods,* 2: 3-14.

Adhikari, B. (2005), "Poverty, Property Rights and Collective Action: Understanding the Distributive Aspects of Common Property Resource Management", *Environment and Development Economics* 10: 7-31.

Adhikari, B., S. di Falco and J.C. Lovett (2004), "Household Characteristics and Forest Dependency: Evidence from Common Property Forest Management in Nepal", *Ecological Economics,* 48:245 257.

Aggarwal, R.M. and T.A. Narayan (2004), "Does Inequality Lead to Greater Efficiency in the Use of Local Commons? The Role of Strategic Investments in Capacity", *Journal of Environmental Economics and Management* 47, 163-182.

Alavalapati, J.R.R., W.L. Adamowicz and W.A. White (1999), "Distributive Impacts of Forest Resource Policies in Alberta", *Forest Science* 45(3), 342-348.

Albers H.J. and E. Grinspoon (1997), "A Comparison of the Enforcement of Access Restrictions Between Xishuangbanna Nature Reserve (China) and Khao Yai National Park (Thailand)", *Environ. Conserv.* 24:351-62.

Aldred, J. and M. Jacobs (2000), "Citizens and Wetlands: Evaluating the Ely Citizens' Jury", *Ecological Economics,* 34:217 232.

Alexander, J. and J.-A. McGregor (2000), "Wildlife and Politics: CAMPFIRE in Zimbabwe", *Development and Change* 31(3), 605-627.

Alix-Garcia, J., A. de Janvry and E. Sadoulet (2004), "A Tale of Two Communities: Explaining Deforestation in Mexico", *World Development* 33(2), 219-235.

Allali-Puz H., E. Béchaux and C. Jenkins (2003), "Governance et democratic locale dans les Parcs Naturels Régionaux de France", *Policy Matters* 12:225-237.

Allegretti, M. (1990), "Extractive Reserves: An Alternative for Reconciling Development and Environmental Conservation in Amazonia", in Anderson, A. (ed.) (1990) *Alternatives to Deforestation: Steps Toward Sustainable Use of the Amazon Rain Forest,* Columbia University Press, New York.

Allegretti, M. (2002), *A construção social de políticas ambientais: Chico Mendes e o Movimento dos Seringueiro,* Centro de Desenvolvimento Sustentável, Universidade de Brasília PhD Thesis, Brasília, Brazil.

Allgood, S. and A. Snow (1998), "The Marginal Cost of Raising Tax Revenue and Redistributing Income", *Journal of Political Economy* 106(6), 1246-1273.

Alston, L. *et al.* (1999), "A model of rural conflict: violence and land reform policy in Brazil", *Environment and Development Economics* 4, 135-160.

Amend, S. and T. Amend (1995), *National Parks Without People? The South American Experience*, IUCN, Gland, Switzerland.

Amiel, Y., J. Creedy and S. Hurn (1999), "Measuring Inequality Aversion", *Scandinavian Journal of Economics* 101 (1), 83-96.

Andersen, I.-E. and B. Jaeger (1999), "Danish Participatory Models: Scenario Workshops and Consensus Conferences: Towards More Democratic Decision-making", *Science and Public Policy*, 5: 331-340.

Angelsen, A., and S. Wunder (2003), *Exploring the Forest-Poverty Link: Key Concepts, Issues and Research Implications*, Center for International Forestry Research, Bogor, Indonesia.

Arnot, C., P. Boxall and S.B. Cash (2006), "Do Ethical Consumers Care About Price? A Revealed Preference Analysis of Fair Trade Coffee Purchases", *Canadian Journal of Agricultural Economics/Revue canadienne d'agroéconomie* 54 (4), 555-565.

Arrow, K.J. (1950), "A Difficulty in the Concept of Social Welfare", *Journal of Political Economy* 58(4) (August, 1950), 328-346.

Asheim, G.B., W. Buchholz and B. Tungodden (2001), "Justifying Sustainability", *Journal of Environmental Economics and Management* 41(3), 252-268.

Atkinson, A. and F. Bourguignon (1982), "The Comparison of Multi-Dimensioned Distributions of Economic Status", *Review of Economic Studies* 49 (1982), 183-201.

Atkinson, A.B. (1970), "On the Measurement of Inequality", *Journal of Economic Theory* 2, 244-263.

Baland, J.-M. and J.-P. Platteau (1997), "Wealth Inequality and Efficiency in the Commons Part I: The Unregulated Case", *Oxford Economic Papers* 49, 451-482.

Baland, J.-M. and J.-P. Platteau (1998), "Wealth Inequality and Efficiency in the Commons Part II: The Regulated Case", *Oxford Economic Papers* 50, 1-22.

Balmford, A. *et al.* (2000), "Integrating Conservation Costs into International Priority Setting", *Conservation Biology* 11, 597-605.

Balmford, A. *et al.* (2001), "Conservation Conflicts Across Africa", *Science* 291 (30 March), 2616-2619.

Balmford, A., *et al.* (2003), "Global Variation in Terrestrial Conservation Costs, Conservation Benefits, and Unmet Conservation Needs", *Proceedings of the National Academy of Sciences of the United States of America* 100, 1046-1050.

Balmford, A. and T. Whitten (2003), "Who Should Pay for Tropical Conservation, and How Could the Costs be Met?" *Oryx* 37, 238-250.

Bannon, I. and P. Collier (2003), "Natural Resources and Conflict: What We Can Do", in *Natural Resources and Violent Conflict: Options and Actions*, World Bank, Washington, DC.

Barbier, E.B. and M. Cox (2004), "An Economic Analysis of Shrimp Farm Expansion and Mangrove Conservation in Thailand", *Land Economics* 80(3), 389-407.

Barbier, E.B., and M. Rauscher (1995), "Policies to Control Tropical Deforestation: Trade Intervention *versus* Transfers", in C. Perring *et al.* (ed.), *Biodiversity Loss: Economic and Ecological Issues*, Cambridge University Press, Cambridge.

Bardhan. P. (1996), "Efficiency, Equity and Poverty Alleviation: Policy Issues in Less Developed Countries", *Economic Journal* 106, 1344-1356.

Barrett, C.B., D.R. Lee and J.G. McPeak, (2005), "Institutional Arrangements for Rural Poverty Reduction and Resource Conservation", *World Development*, Vol. 33(2), 193-197.

Baumol, W.J. and W.E. Oates (1988), *The Theory of Environmental Policy*, Cambridge University Press, Cambridge.

Bedunah D.J. and S.M. Schmidt (2004), "Pastoralism and Protected Area Management in Mongolia's Gobi Gurvansaikhan National Park", *Dev. Change* 35(1): 167-91.

Bellon, M.R. and J.E. Taylor (1993), "Folk Soil Taxonomy and the Partial Adoption of New Seed Varieties", *Economic Development and Cultural Change*, 41(4), 763-786.

Bergstrom, T.C. and R.P. Goodman (1973), "Private Demands for Public Goods", *American Economic Review*, 63(3), 280-296.

Bergstrom, T., L. Blume and H. Varian (1986), "On the Private Provision of Public Goods", *Journal of Public Economics* 29, 25-49.

Berkes, F. (1999), *Sacred Ecology: Traditional Ecological Knowledge and Resource Management*, Taylor and Francis, Philadelphia, USA.

Beukering, P.H. van, H. Cesara and M.A. Janssen (2003), "Economic Valuation of the Leuser National Park on Sumatra, Indonesia", *Ecological Economics* 44(1), February 2003, 43-62.

Bingham, G. (1986), *Resolving Environmental Disputes, A Decade of Experience*, The Conservation Foundation, Washington DC.

Bojo, J. and R.C. Reddy (2002), *Poverty Reduction Strategies and Environment: A Review of 40 Interim and Full Poverty Reduction Strategy Papers*, World Bank, Washington D.C.

Borcherding, T.E. and R.T. Deacon (1972), "Demand for Services of Non-Federal Governments", *American Economic Review*, 62(5), 891-901.

Borrini-Feyerabend, G. *et al.* (2004), *Sharing Power: Learning by Doing in Co-management of Natural Resources Throughout the World*, IIED and IUCN/CEESP/CMWG, Cenesta, Tehran.

Bovenberg, A.L. and B.J. Heijdra (1998), "Environmental Tax Policy and Intergenerational Distribution", *Journal of Public Economics* 67, 1-24.

Boyce, J.K. (2002), *The Political Economy of the Environment*, Edward Elgar, Cheltenham, UK

Brainard, J.S. *et al.* (2006), "Exposure to Environmental Urban Noise Pollution in Birmingham, UK", in: Serret and Johnstone (eds.), *The Distributional Effects of Environmental Policy*, Edward Elgar, Cheltenham, UK.

Brett, C. and M. Keen (2000), "Political Uncertainty and the Earmarking of Environmental Taxes", *Journal of Public Economics* 75, 315-340.

Brooks, N and R. Sethi (1997), "The Distribution of Pollution: Community Characteristics and Exposure to Air Toxics", *Journal of Environmental Economics and Management*, 32, 233-250.

Broome, J. (1992), *Counting the Cost of Global Warming*, White Horse Press, Cambridge.

Brown, K. (1998), "The Political Ecology of Biodiversity, Conservation and Development in Nepal's Terai: Confused Meanings, Means and Ends", *Ecological Economics* 24(1), 73-87.

Brown, K. and S. Rosendo (2000), "Environmentalists, Rubber Tappers and Empowerment: The Politics and Economics of Extractive Reserves", *Development and Change*, 31: 201-227.

Brown, K. *et al.* (2001), "Trade-off Analysis for Marine Protected Area Management", *Ecological Economics*, 37: 417-434.

Bruner A. *et al.* (2001), "Effectiveness of Parks in Protecting Tropical Biodiversity", *Science* 291(5501): 125-28.

Buchanan, J.M. (1963), "The Economics of Earmarked Taxes", *Journal of Political Economy* 71(5), 457-469.

Buchy, M., H. Ross and W. Proctor (2000), *Enhancing the Information Base on Participatory Approaches in Australian Natural Resources Management*, Commissioned Report to the Land and Water Resources Research and Development Corporation, Canberra.

Bueno de Mesquita, B. *et al.* (2003), *The Logic of Political Survival*, MIT Press, Cambridge, Mass.

Bulte, E. and C. Withagen (2006), *Distributive Issues in a Dynamic Context: an Issues Paper*, OECD, Paris.

Bulte, E.H., R. Damania and R.T. Deacon (2005), "Resource Intensity, Institutions, and Development", *WorldDevelopment* 33(7), 1029-1044.

Burnham, P. (2000), *Indian Country God's Country: Native Americans and National Parks*, Island Press, Washington, DC.

Burton, P.S. (2004), "Hugging Trees: Claiming *de facto* Property Rights by Blockading Resource Use", *Environmental and Resource Economics* 27, 135-163.

Campbell, B. *et al.* (2001), "Challenges to Proponents of Common Property Resource Systems: Despairing Voices from the Social Forests of Zimbabwe", *World Development* 29: 589-600.

Canadian Model Forest Network (2006), *Canadian Model Forest Network: Achievements*, Natural Resources Canada, Ottawa.

Carruthers J. (1995), *The Kruger National Park: A Social and Political History*, Univ. Natal Press, Pietermaritzburg, South Africa.

Carson, L. and K. Gelber (2001), *Ideas for Community Consultation: A Discussion on Principles and Procedures for Making Consultation Work*, NSW Department of Urban Affairs and Planning, Sydney, Australia.

Catton T. (1997), *Inhabited Wilderness: Indians, Eskimos, and National Parks in Alaska*, Univ. N. Mex. Press, Albuquerque.

Cavendish, W. (2000), "Empirical Regularities in the Poverty-Environment Relationship of Rural Households: Evidence from Zimbabwe", *World Development*, 28, (11), 1979-2003.

CBD (Convention on Biological Diversity) (1992), *Convention on Biological Diversity*, *http://sedac.ciesin.org/entri/texts/biodiversity.1992.html*.

CBD (2005), *Thematic Report on Mountain Ecosystems, Nepal*, *www.biodiv.org/doc/world/np/np-nr-me-en.doc*.

 PEOPLE AND BIODIVERSITY POLICIES – ISBN 978-92-64-03431-0 –

Cernea, M.M. and K. Schmidt-Soltau (2006), "Poverty Risks and National Parks: Policy Issues in Conservation and Resettlement", *World Development* 34(10), 1808-1830.

Chakraborty, R.N. (2001), "Stability and Outcomes of Common Property Institutions in Forestry: Evidence from the Terai Region of Nepal", *Ecological Economics* 36, 341-353.

Chapin, M. (2004), "A Challenge to Conservationists", *World Watch Magazine*, November/December 2004, 17-31.

Chatty, D. and M. Colchester (eds.) (2002), *Conservation and Mobile Indigenous Peoples: Displacement, Forced Settlement and Sustainable Development*, Berghahn Books, New York.

Chichilinsky, G. (1996), "An Axiomatic Approach to Sustainable Development", *Social Choice and Welfare* 13, 231-257.

Chichilnisky, G. and G. Heal (1994), "Who Should Abate Carbon Emissions? An International Viewpoint", *Economics Letters* 44, 443-449.

Chobotova, V. and T. Kluvankova-Oravska (2006), *Community-based Management of Biodiversity Conservation in a Transition Economy. Application of Multi-Criteria Decision Aid to the Nature Reserve Šúr*, case study prepared for OECD, OECD, Paris.

Clark, C.W. (1973), "Profit Maximization and the Extinction of Animal Species", *Journal of Political Economy* 81(4), 950-961.

Clean Water Action Plan (2000), *Watershed Success Stories: Applying the Principles and Spirit of the Clean Water Action Plan*, USA

Cleary, D. (2006), "The Questionable Effectiveness of Science Spending by International Conservation Organizations in the Tropics", *Conservation Biology* 20(3), 733-738.

Clippel, G. de (2005), *Equity, Envy, and Efficiency under Asymmetric Information*, Working Paper, Rice University, Houston.

Cobham, A. (2007), *Tax Evasion, Tax Avoidance and Development Finance*, University of Oxford, Department of International Development, Oxford.

Coomes, O., B. Barham, and Y. Takasaki (2004), "Targeting Conservation-Development Iniatives in Tropical Forests: Insights from Analysis of RainForest Use and Economic Reliance among Amazonian peasants", *World Development* 55, 47-64.

Cooperative Conservation America (2005), *Faces and Places of Cooperative Conservation*, report of White House Conference on Cooperative Conservation, St. Louis, Missouri, August 29-31, 2005, US Department of the Interior, Washington DC.

Cork, S. (2002), "What are Ecosystem Services?", *RIPRAP* (River and Riparian Lands Management Newsletter), Land and Water Australia, Canberra, 21, pp.1-9.

Costanza, R. *et al.* (1997), "The Value of the World's Ecosystem Services and Natural Capital", *Nature* 387, 253-261.

Cowell, F.A. and K. Gardiner (1999), "Welfare Weights", *STICERD, London School of Economics, Economics Research Paper* 20, Aug 1999, LSE, London.

Crosby, N. (1996), *Creating an Authentic Voice of the People: Deliberation on Democratic Theory and Practice*. Midwest Political Science Association, Chicago, USA.

CSIRO (Commonwealth Scientific and Industrial Research Organisation) (2003), *Natural Values: Exploring Options for Enhancing Ecosystem Services in the Goulburn Broken Catchment*, Ecosystem Services Project, CSIRO, Canberra, Australia.

Dasgupta, P. (2000), "Valuing Biodiversity", in Levin, S. (ed.) *Encyclopedia of Biodiversity*, Academic Press, New York.

Datta, S.K. and S. Kapoor (1996), *Collective Action, Leadership and Success in Agricultural Cooperatives – a Study of Gujarat and West Bengal*, Oxford and IBH Publishing, Oxford and New Dehli.

Day-Rubinstein, K. and G.B. Frisvold (2001), "Genetic Prospecting and Biodiversity Development Agreements", *Land Use Policy* 18(3), 205-219.

Deacon, R.T. (2006), "Distributive Issues Related to Biodiversity: The Role of Institutions", presentation prepared for the *OECD Workshop on Distributive Issues Related to Biodiversity*, Oaxaca, Mexico, April 26-27, 2006.

Declerck, S. (2003), "Restoration of Lake Kraenepoel in Belgium, a Case Study Prepared for the BIOFORUM Project", in: Young, J. *et al.* (eds.), *Conflicts Between Human Activities and the Conservation of Biodiversity in Agricultural Landscapes, Grasslands, Forests, Wetlands and Uplands in Europe*, Report of the BIOFORUM projects, August, 2003, 116-119, BIOFORUM, Centre for Ecology and Hydrology, Edinburgh.

Demsetz, H. (1967), "Toward a Theory of Property Rights", *American Economic Review* 57(2), Papers and Proceedings, 347-359.

Department of Sustainability and Environment (DSE) (2005a), *Southern Victoria BushTender: Information Sheet* No. 5, Victorian Government Department of Sustainability and Environment, Melbourne.

DSE (2005b), *Southern Victoria BushTender: Information Sheet No. 6*, Victorian Government Department of Sustainability and Environment, Melbourne.

DSE (2005c), *Southern Victoria BushTender: Information Sheet No. 7*, Victorian Government Department of Sustainability and Environment, Melbourne.

Diamond, J. (2005), *Collapse: How Societies Choose to Fail or Succeed*, Viking, New York.

Dietz, T., E. Ostrom and P.C. Stern (2003), "The Struggle to Govern the Commons", *Science* 302, 1907-1912.

Dixit, A.K. and J.E. Stiglitz (1977), "Monopolistic Competition and Optimum Product Diversity", *American Economic Review*, 67(3), 297-308.

Dixon, J.A. and P.B. Sherman (1990), *Economics of Protected Areas: A New Look at Benefits and Costs*, East-West-Center Center, Island Press, Washington DC.

Dixon, J.A. and P.B. Sherman (1991), "Economics of Protected Areas", *Ambio*, 20(2), 68-74.

Drazen, A. (2001), *Political Economy in Macroeconomics*, Princeton University Press, Princeton.

Drechsler, M. *et al.* (2007), "An Agglomeration Payment for Cost-Effective Biodiversity Conservation in Spatially Structured Landscapes", *UFZ Discussion Papers* 4/2007, March 2007, UFZ Centre for Environmental Research Leipzig, Germany.

Dressler, W.H. (2006), "Co-opting Conservation: Migrant Resource Control and Access to National Park Management in the Philippine Uplands", *Development and Chance* 37(2), 401-426.

Drèze, J.P. (1998), "Distribution Matters in Cost-Benefit Analysis: Comment on K-A. Brekke", *Journal of Public Economics* 70 (3): 485-88.

Drèze, J.P. and N. Stern (1987), "The Theory of Cost-Benefit Analysis", in A.J. Auerbach and M. Feldstein (eds.) *Handbook of Public Economics* 2, North-Holland, Amsterdam.

Droege, S. and B. Soete (2001), "Trade-Related Intellectual Property Rights, North-South Trade and Biological Diversity", *Environmental and Resource Economics* 19, 149-163.

Dublin, H., C. Volonte and J. Brann (2004), *GEF Biodiversity Program Study*, Washington, D.C.: Monitoring and Evaluation Unit, Global Environment Facility Secretariat.

Easterbrook, G. (2003), *The Progress Paradox*, Random House, New York.

Emerton, L., J. Bishop and L. Thomas (2005), *Sustainable Financing of Protected Areas: A Global Review of Challenges and Options*, IUCN, Gland, Switzerland and Cambridge, UK.

Engel, S., R. Lopez and C. Palmer (2006), "Community–Industry Contracting over Natural Resource use in a Context of Weak Property Rights: The Case of Indonesia", *Environmental and Resource Economics* 33(1), 73-93.

Environment Canada (2005), *The Canadian Ecological Gifts Program Handbook 2005: A Legacy for Tomorrow, a Tax Break Today*, available at: *www.cws-scf.ec.gc.ca/ecogifts/hb_toc_e.cfm*.

Environmental Defense (2000), *Progress on the Back Forty: An Analysis of the Three Incentive Based Approaches to Endangered Species Conservation on Private Lands*, Environmental Defense, New York.

EPA (US Environmental Protection Agency) (2001), *Protecting and Restoring America's Watersheds: Status, Trends, and Initiatives in Watershed Management*, EPA-840-R-00-001, US EPA, Washington DC.

Eskeland, G. and C. Kong (1998), "Protecting the Environment and the Poor: A Public Goods Framework Applied to Indonesia", *World Bank Policy Research Working Paper* No. 1961, World Bank, Washington, DC.

European Commission (2005), *Agri-environment Measures: Overview on General Principles, Types of Measures, and Application*, study of the European Commission Directorate General for Agriculture and Rural Development, Unit G-4, Evaluation of Measures applied to Agriculture, available at: *http://ec.europa.eu/agriculture/publi/reports/agrienv/rep_en.pdf*.

Fearnside, P.M. (2003), "Conservation Policy in Brazilian Amazonia: Understanding the Dilemmas", *World Development* 31(5): 757-779.

Feinerman, E., A. Fleischer and A. Simhon (2004), "Distributional Welfare Impacts of Public Spending: The Case of Urban *versus* National Parks", *Journal of Agricultural and Resource Economics* 29(2): 370-386.

Ferraro, P.J. (2002), "The Local Costs of Establishing Protected Areas in Low-Income Nations: Ranomafana National Park, Madagascar", *Ecological Economics*, 43: 261-275.

Ferraro, P.J. and D. Simpson (2002), "The Cost-Effectiveness of Conservation Payments", *Land Economics* 78(3), 339-353.

Fisher, M. (2004), "Household Welfare and Forest Dependence in Southern Malawi", *Environment and Development Economics* 9: 135-154.

Fisher, M., G.E. Shively and S. Buccola (2005), "Activity Choice, Labor Allocation, and Forest Use in Malawi", *Land Economics* 81 (4), 503-517.

Fisher, R., W. Ury and B. Patton (1991), *Getting to Yes: Negotiating Agreement Without Giving In*, Penguin Books, New York.

Fishkin, J. and R.C. Luskin (2004), "Experimenting with a Democratic Ideal: Deliberative Polling and Public Opinions", paper prepared for presentation at the *Swiss Chair's Conference on Deliberation*, The European University Institute, Florence, Italy, May 21-22, 2004.

Flores, N. and R. Carson (1997), "The Relationship Between the Income Elasticities of Demand and Willingness to Pay", *Journal of Environmental Economics and Management* 33, 287-295.

Fraga, J. (2006), "Local Perspectives In Conservation Politics: The Case of the Ria Lagartos Biosphere Reserve, Yucatan, Mexico", Landscape and Urban Planning, 74(3-4), 285-295

Frank, G. and F. Müller (2003), "Voluntary Approaches in Protection of Forests in Austria", *Environmental Science and Policy*, 6: 261-269.

Frederick, S., G. Loewenstein and T. O'Donoghue (2002), "Time Discounting and Time Preferences: A Critical Review", *Journal of Economic Literature* 40, 351-401.

Freudenburg, W., L. Wilson and D. O'Leary (1998), "Forty Years of Spotted Owls? A Longitudinal Analysis of Logging Industry Job Losses", *Sociological Perspectives* 41(1), 1-26.

Gale, D. (1973), "Pure Exchange Equilibrium In Dynamic Economic Models", *Journal of Economic Theory* 6, 12-36.

Gaston, K. (2005), "Biodiversity and Extinction: Species and People", *Progress in Physical Geography* 29(2), 239-247.

Gatti, R. *et al.* (2004), "The Biodiversity Bargaining Problem", *Cambridge Working Papers in Economics*, No. 0447, Faculty of Economics, University of Cambridge, Cambridge, UK.

Gawler, M. (ed.) (2002), "Strategies for Wise Use of Wetlands: Best Practices in Participatory Management", in proceedings of a workshop held at the *2nd International Conference on Wetlands and Developments* (November 1998, Dakar, Senegal), Wetlands International, IUCN, WWF publication No. 56, Wageningen, Netherlands.

GEF (Global Environment Facility), 2006, "The Role of Local Benefits in Global Environmental Programs", *Evaluation Report No.* 30, Global Environment Facility Evaluation Office, Washington DC.

Geisler, C. and de Sousa, R. (2001), "From Refuge to Refugee: The African Case", *Public Adm. Dev.* 21: 159-70.

Gerlagh, R. and M.A. Keyzer (2001), "Sustainability and the Intergenerational Distribution of Natural Resource Entitlements", *J. Public Econom.* 79 (2001) 315-341.

Gibson, C.C., J.T. Williams and E. Ostrom (2005), "Local Enforcement and Better Forests", *World Development* 33(2), 273-284.

Gjertsen, H. (2005), "Can Habitat Protection Lead to Improvements in Human Well-Being? Evidence from Marine Protected Areas in the Philippines", *World Development* 33(2), 199-217.

Gjertsen, H. and C.B. Barrett (2004), "Context-Dependent Biodiversity Conservation Management Regimes: Theory and Simulation", *Land Economics* 80(3): 321-339.

Goeschl, T. and D. Igliori (2004), "Reconciling Conservation and Development: A Dynamic Hotelling Model of Extractive Reserves", *Land Economics* 80(3), 340-354.

Goeschl, T. and D. Igliori (2006), "Property Rights for Biodiversity Conservation and Development: Extractive Reserves in the Brazilian Amazon", *Development and Change* 37(2), 427-51.

Gollier, C. (2002a), "Time Horizon and the Discount Rate", *Journal of Economic Theory* 107(2), 463-473.

Gollier, C. (2002b), "Discounting an Uncertain Future", *Journal of Public Economics* 85, 149-166.

Googch, G.D., G. Jansson and R. Mikaelsson (2003), *Results of Focus Groups Conducted in the River Basin Area of Motala Ström, Sweden*, River Dialogue Project, Department of Management and Economics, Political Science, Linköping University.

Grady, S. (2000), "Kakadu National Park, Australia, Case study 11", in Beltran, J. (ed.), *Indigenous and Traditional Peoples and Protected Areas: Principles, Guidelines and Case Studies*, IUCN, Gland, Switzerland.

Grimble, R. *et al.* (1995), "Trees and Trade-Offs: A Stakeholder Approach to Natural Resource Management", *Gatekeeper Series* No. 52., International Institute for Environment and Development, London.

Groier, M. (2004), "Socioeconomic effects of the Austrian Agro-Environmental Program. Mid-Term Evaluation 2003", *Facts and Feature* 27. Bundesanstalt für Bergbauernfragen, Vienna.

Groom, B., *et al.* (2005), "Declining Discount Rates: The Long and the Short of it", *Environmental and Resource Economics* 32(4), 445-493.

Hamilton, J.T. (2006), "Environmental Equity and the Sitting of Hazardous Waste Facilities in OECD Countries", in Serret and Johnstone (eds.), *The Distributional Effects of Environmental Policy*, Edward Elgar, Cheltenham, UK.

Hanley, N. and C. Spash (1993), *Cost Benefit Analysis and the Environment*, Edward Elgar, Cheltenham.

Hardin, G. (1968), "The Tragedy of the Commons", *Science* 168(3859), Dec. 13th 1968, 1243-48.

Harford, T. (2003), "Fair Trade Coffee Has a Commercial Blend", *Financial Times*, 12 Sept. 2003, 15.

Haro, G.O., G.J. Doyo and J.G. McPeak (2005), "Linkages Between Community, Environmental, and Conflict Management: Experiences from Northern Kenya", *World Development* 33(2), 285-299.

Heady, C. (2000), "Natural Resource Sustainability and Poverty Reduction", *Environment and Development Economics*, 5: 241-258.

Heal, G. (1999), "Markets and Sustainability", *The Science of The Total Environment* 240(1-3), October 1999, 75-89.

Hegan, R.L., G. Hauer and M.K. Luckert (2003), "Is the Tragedy of the Commons Likely? Factors Preventing the Dissipation of Fuelwood Rents in Zimbabwe", *Land Economics* 79 (2): 181-197.

Hepburn, C. (2006), "*Use of Discount Rates in the Estimation of the Costs of Inaction with Respect to Selected Environmental Concerns*", Working Party on National Environmental Policies, OECD, Paris.

Herrera, A. and da Passano, M.G. (2006), "Land Tenure Alternative Conflict Management", *FAO Land Tenure Manuals* No. 2, Food and Agriculture Organisation of the United Nations, Land Tenure Service, Rural Development Division, Rome.

Hiedanpää, J. (2002), "European-Wide Conservation Versus Local Well-Being: The Reception of the Natura 2000 Reserve Network in Karvia, SW-Finland", *Landscape and Urban Planning* 61: 113-123.

HM Treasury (2003), *The Green Book – Appraisal and Evaluation in Central Government – Treasury Guidance*, TSO, London.

Hökby, S. and T. Söderqvist (2003), "Elasticities of Demand and Willingness to Pay for Environmental Services in Sweden", *Environmental and Resource Economics*, 26, 361-383.

Homma, A.K.O. (1992), "The Dynamics of Extraction in Amazonia: A Historical Perspective", in Nepstad, D.C. and S. Schwartzman (eds.), *Non-Timber Products from Tropical Forests: Evaluation of a Conservation and Development Strategy*, Advances in Economic Botany 9: 33-42, The New York Botanical Garden, New York.

Horne, P. (2004), "Forest Owners' Acceptance of Incentive Based Instruments in Forest Biodiversity Conservation – A Choice Experiment Based Approach", paper presented at the *48th Annual Conference of the Australian Agriculture and Resource Economics Society*.

Horne, P. and A. Naskali (2006), *Voluntary Scheme for Forest Protection on Private Land as Part of the METSO Programme in Finland*, Finnish Forest Research Institute, case study prepared for OECD, Paris.

Horowitz, J.K. and K.E. McConnell (2003), "Willingness to Accept, Willingness to Pay and the Income Effect", *Journal of Economic Behavior and Organization*, 51(4), 537-545.

Horton, B., *et al.* (2003), "Evaluating Non-Users' Willingness to Pay for the Implementation of a Proposed National Parks Program in Amazonia", *Environmental Conservation* 20(2), 139-146.

Howarth, R. (2000), "*Normative Criteria for Climate Change Policy Analysis*", Redefining Progress, San Francisco.

Hubacek, K and W. Bauer (1999), *Economic Incentive Measures in the Creation of the National Park Neusiedler See Seewinkel*, OECD, Paris.

Humphreys, D. (2001), "Forest Negotiations at the United Nations: Explaining Cooperation and Discord", *Forest Policy and Economics*, 3: 125-135.

Islam, M and J.B. Braden (2006), "Bio-economic Development of Floodplains: Farming Versus Fishing in Bangladesh", *Environment and Development Economics* 11, 95-126.

James, R.F. (1999), "Public Participation and Environmental Decision-Making – New Approaches", paper presented at the *National Conference of the Environmental Institute of Australia*, 1-3 December, 1999.

James, R.F. and R.K. Blamey (2000), *A Citizens' Jury Study of National Park Management, Canberra*, Australian National University, Canberra, available at: *http://cjp.anu.edu.au*.

Jepson, P., F. Momberg and H. van Noord (2002), "A Review of the Efficacy of the Protected Area System of East Kalimantan Province, Indonesia", *Nat. Areas J.* 22(1): 28-42.

Johannesen, A.B. and A. Skonhoft (2004), "Property Rights and Natural Resource Conservation. A Bio-Economic Model with Numerical Illustrations from the Serengeti-Mara Ecosystem", *Environmental and Resource Economics* 28(4), 469-488.

Johansson-Stenman, O. (2005), "Distributive Weights in Cost-Benefit Analysis – Should We Forget About Them?", *Land Economics* 81(3), 337-352.

Jones, B. and M. Murphree (2001), "The Evolution of Policy on Community Conservation in Namibia and Zimbabwe", in D. Hulme and M. Murphree (eds.) *African Wildlife and Livelihoods: The Promise and Performance of Community Conservation*, James Currey, Oxford.

Just, R.E., D.L. Hueth and A. Schmitz (2004), *The Welfare Economics of Public Policy*, Edward Elgar, Cheltenham, UK.

Just, R.E. and D.L. Hueth (1979), "Multimarket Welfare Measurement", *American Economic Review* 69(5), 947-54.

Justino, P., J. Litchfield and Y. Niimi (2004), "Multidimensional Inequality: An Empirical Application to Brazil", *PRUS Working Paper* No. 24, Poverty Research Unit, Department of Economics, University of Sussex.

Kahn, M. and J. Matsusaka (1997), "Demand for Environmental Goods. Evidence from Voting Patterns on California Initiatives", *Journal of Law and Economics* 40, 137-173.

Kakwani, N.C. (1977), "Measurement of Tax Progressivity: An International Comparison", *Economic Journal* 87(345), 71-80.

Kalter, R.J. and T.H. Stevens (1971), "Resource Investment, Impact Distribution, and Evaluation Concepts", *American Journal Agricultural Economics*, 53(2), 206-215.

Kelly, B., M. Brown and O. Byers (eds.) (2001), *Mexican Wolf Reintroduction Program,Three-Year Review Workshop: Final Report*, IUCN/SSC Conservation Breeding Specialist Group, Apple Valley, MN, USA.

Kenyon, W. and C. Nevin (2001), "The Use of Economic and Participatory Approaches to Assess Forest Development: A Case Study in the Ettrick Valley", *Forest Policy and Economics* 3: 69-80.

Khare, A. and D. Bray (2004), *Study of Critical New Forest Conservation Issues in the Global South*, Ford Foundation, New York.

Kishor, N. and R. Damania (2006), "Crime and Justice in the Garden of Eden: Improving Governance and Reducing Corruption in the Forestry Sector", in J. Edgardo Campos and S. Pradhan (eds.), *The Many Faces of Corruption: Tracking Vulnerabilities at the Sector Level*, The World Bank, Washington, DC.

Kolm, S. (1977), "Multidimensional Egalitarianisms", *Quarterly Journal of Economics* 91 (1977), 1.

Konisky, D.M. and T.C. Beierle (2001), "Innovation in Public Participation and Environmental Decision Making: Examples from the Great Lakes Region", *Research Note, Society and Natural Resources* 14: 815-826.

Kontogianni A. *et al.* (2001), "Integrating Stakeholder Analysis in Non-Market Valuation of Environmental Assets", *Ecological Economics* 37: 123-138.

Koopmans, T. (1965), "On the Concept of Optimal Economic Growth", in: Pontificiae Academiae Scientiarium Scriptum Varia (ed.): *The Economic Approach to Development Planning*, North-Holland, Amsterdam.

Kothari A. (2004), "Displacement Fears", *Frontline,* 21(26), 18-31 Dec., India. Available at *www.frontlineonnet.com/fl2126/stories/20041231000108500.htm.*

Kooten, G.C. van and E.H. Bulte (2000), *The Economics of Nature: Managing Biological Assets*, Wiley-Blackwell Publishing.

Kramer, R. and E. Mercer (1997), "Valuing a Global Environmental Good: US Residents' Willingness to Pay to Protect Tropical Rain Forests", *Land Economics* 73, 196-210.

Krautkraemer, J.A. and R.G. Batina (1999), "On Sustainability and Intergenerational Transfers with a Renewable Resource", *Land Economics* 75, 167-184.

Kriström, B. (2006), "Framework for Assessing the Distribution of Financial Effects of Environmental Policy", in Y. Serret and N. Johnstone (eds.), *The Distributional Effects of Environmental Policy*, Edward Elgar, Cheltenham, UK.

Kriström, B and P. Riera (1996), "Is the Income Elasticity of Environmental Improvements Less Than One?"*Environmental and Resource Economics*, 7, 45-55.

Krüger O. (2004), "The Role of Ecotourism in Conservation: Panacea or Pandora's Box?"*Biodivers. Conserv.* 14(3): 579-600.

Krutilla, J.V. (1967), "Conservation Reconsidered", *American Economic Review* 57(4), 777-786.

Kumar, S. (2002), "Does Participation' in Common Pool Resource Management Help the Poor? A Social Cost-Benefit Analysis of Joint Forest Management in Jharkhand, India", *World Development* 30: 763-782.

Lake, D. and M. Baum (2001), "The Invisible Hand of Democracy: Political Control and the Provision of Public Services", *Comparative Political Studies* 34(6), 587-621.

Langholz, J.A. and W. Krug (2004), "New Forms of Biodiversity Governance: Non State Actors and the Private Protected Area Action Plan", *Journal of International Wildlife Law and Policy*, 7, 9-29.

Lawrence, D. (2000), *Kakadu: The Making of a National Park*, Melbourne Univ. Press, Melbourne, Australia.

Leakey, R.E., and R. Lewin (1995), *Sixth Extinction: Patterns of Life and the Future of Humankind*, Anchor Books, New York.

Lee, D.R. and C.B. Barrett (2001), *Tradeoffs or Synergies? Agricultural Intensification, Economic Development and the Environment*, CABI Publishing, Wallingford, UK.

Libecap, G.D. and J. Smith (2002), "The Economic Evolution of Petroleum Property Rights in the United States", *Journal of Legal Studies* 31(2), 589-608.

Li, C.Z. and K.G. Löfgren (2000), "Renewable Resources and Economic Sustainability: A Dynamic Analysis with Heterogeneous Time Preferences", *Journal of Environmental Economics and Management* 40, 236-250.

Lind, R.C. (1995), "Intergenerational Equity, Discounting, and the Role of Cost-Benefit Analysis in Evaluating Global Climate Policy", *Energy Policy* 23: 379-389.

Linde-Rahr, M. (1998), *Rural Reforestation: Gender Effects on Private Investments in Vietnam*, Working Paper, Department of Economics, Goteborg University, Sweden.

Lopez, T.T. de (2003), "Economics and Stakeholders of Ream National Park, Cambodia", *Ecological Economics* 46: 269-282.

Luck, G. *et al.* (2004), "Alleviating Spatial Conflict Between People and Biodiversity", *Proceedings of the National Academy of Sciences* 101(1), 182-186.

Lusty, C. (2000), "The Lapponian Area, Sweden", Case study 5, in Beltran, J. (ed.), *Indigenous and Traditional Peoples and Protected Areas: Principles, Guidelines and Case Studies*, IUCN, Gland, Switzerland.

Lybbert, T.J., C.B. Barrett and H. Narjisse (2002), "Market-based Conservation and Local Benefits: The Case of Argan Oil in Morocco", *Ecological Economics* 41, 125-144.

Lynch, L. and S. Lovell (2003), "Combining Spatial and Survey Data to Explain Participation in Agricultural Land Preservation Programs", *Land Economics* 79 (2): 259-276.

Maasoumi, E. (1986), "The Measurement and Decomposition of Multi-Dimensional Inequality", *Econometrica* 54 (1986), 991-997.

Mahatny S. and D. Russel (2002), "High Staked: Lessons from Stakeholder Groups in the Biodiversity Conservation Network", *Society and Natural Resources*, 15: 179-188.

Maikhuri, R.K. *et al.* (2000), Analysis and Resolution of Protected Area-People Conflicts in Nanda Devi Biosphere Reserve, India, *Environmental Conservation* 27(1): 43-53.

Marcouiller, D.W. and J.C. Stier (1996), *Modelling the Regional Economic Aspects of Forest Management Alternatives*, research paper, McIntere Stennis Program of USDA, University of Wisconsin, Medison, USA.

Margulis, S. (2004), "Causes of Deforestation of the Brazilian Amazon", *World Bank Working Paper* No. 22, The World Bank, Washington DC.

Markandya, A. (2001), "Poverty Alleviation and Sustainable Development: Implications for the Management of Natural Capital", prepared for the International Institute for Sustainable Development (IISD) *Workshop on Poverty and Sustainable Development*, 23rd January, Ottawa.

Marsiliani, L. and T.I. Renström (2000), "Time Inconsistency in Environmental Policy: Tax Earmarking as a Commitment Solution", *Economic Journal* 110, 123-138.

Mashinya, J. (2007), *Participation and Devolution in Zimbabwe's CAMPFIRE Program: Findings from Local Projects in Mahenyeand Nyamiyami*, Faculty of Graduate School of the University of Maryland, USA.

McLean, J. and S. Straede (2003), "Conservation, Relocation and the Paradigms of Park and People Management – A Case Study of Padampur Villages and the Royal Chitwan National Park, Nepal", *Soc. Nat. Res.* 16: 509-26.

McNeely, J.A. and S.J. Scherr (2003), *Ecoagriculture: Strategies to Feed the World and Save Wild Biodiversity*, Island Press, Washington, DC.

Menezes, M. (1994), "As Reservas Extrativistas como Alternativa ao Desmatamento na Amazônia", in Arnt, R. (ed.) *O Destino da Floresta: Reservas Extrativistas e Desenvolvimento Sustentável na Amazônia*, Relume Dumará, Rio de Janeiro.

Meyer, S. (2001), "Community Politics and Endangered Species Protection", in: Shogren, J. and J. Tschirhart (eds.), *Protecting Endangered Species in the United States. Biological Needs, Political Realities, Economic Choices* Cambridge University Press, Cambridge.

Millimet, D. and D. Slottje (2000), *The Distribution of Pollution in the United States: An Environmental Gini Approach*, working paper, Southern Methodist University, Dallas, Texas.

Mirrlees, J. (1979), *The Implications of Moral Hazard for Optimal Insurance*, mimeo, seminar given at the conference held in honor of Karl Borch, Bergen, Norway.

Moore, C. (1996), *The Mediation Process – Practical Strategies for Resolving Conflict*, 2nd edition, Wiley/Jossey-Bass publishers, San Francisco.

Moore, L, L. Michaelson and S. Orenstein (2000), *Designation of Critical Habitat National Project, Digest of the Process and Results*, Institute of Environmental Conflict Resolution, Tuscon, Arizona.

Morris, C. (2004), "Networks of Agrienvironmental Policy Implementation: A Case Study of England's Countryside Stewardship Scheme", *Land Use Policy*, 21: 177-191.

Mourmouras, A. (1993), "Conservationist Government Policies and Intergenerational Equity in an Overlapping Generations Model with Renewable Resources", *Journal of Public Economics* 51, 249-268.

Mowat, S. (2006), *The Design and Implementation of the Entry Level Scheme in England, DEFRA, UK*, case prepared for the OECD.

Musgrave, R.A. (1959), *The Theory of Public Finance*, McGraw Hill, New York.

Musinguzi, M. (2006), "Making Partnerships for Sustainable Gorilla Tourism in Mgahinga Mountain", *Mountain Forum Bulletin*, Volume VI, Issue 1, January 2006, pp. 4-5 *www.mtnforum.org*.

Naidoo, R. and W.L. Adamowicz (2005), "Biodiversity and Nature-Based Tourism at Forest Reserves in Uganda", *Environment and Development Economics* 10: 159-178.

Naidoo, R. and W.L. Adamowicz (2006a), "Mapping the Economic Costs and Benefits of Conservation", *Public Library of Science-Biology* 4(11), 2153-2163.

Naidoo, R. and W.L. Adamowicz (2006b), "Modeling Opportunity Costs of Conservation in Transitional Landscapes", *Conservation Biology* 20, 490-500.

Nath, S.K. (1969), *A Reappraisal of Welfare Economics*, Routledge, London.

National Round Table on the Environment and the Economy (2005), *Boreal Futures: Governance, Conservation and Development in Canada's Boreal*, National Round Table on the Environment and the Economy, Ottawa.

Natural Resources Canada (2005), *First Nations Forestry Program – Success Stories*, Natural Resources Canada, Canadian Forestry Service, Ottawa (online: *www.fnfp.gc.ca/index_e.php*)

Neary, J.P. (1999), "Comment on Venables (1999) Economic Policy and the Manufacturing Base: Hysteresis in Location", In: Baldwin, R. E., Francois, J. F. (eds.), *Dynamic Issues in Commercial Policy Analysis*, Cambridge University Press, Cambridge, 196-200.

Nepal S..J. (2000), "Wood Buffalo National Park, Canada", Case study 4, in Beltran, J. (ed.), *Indigenous and Traditional Peoples and Protected Areas: Principles, Guidelines and Case Studies*, IUCN, Gland, Switzerland.

Neumann, R. (2004), "Moral and Discursive Geographies in the War for Biodiversity in Africa", *Polit. Geogr.* 23: 813-37.

Nijkamp, P., P. Rietveld and H. Voogd (1990), *Multi-criteria Evaluation in Physical Planning*, North Holland, Amsterdam.

North, D.C. (1990), *Institutions, Institutional Change and Economic Performance*, Cambridge University Press, Cambridge.

O'Connor, M. (2000), "The VALSE project – an introduction", *Ecological Economics* 34: 165-174.

O'Leary, R. and L. Bingham (2004), *The Promise and Performance of Environmental Conflict Resolution*, Resources for the Future, Washington DC.

OECD (Organisation for Economic Co-operation and Development) (1996), *Saving Biological Diversity: Economic Incentives*, OECD, Paris

OECD (1997), *Evaluating Economic Instruments for Environmental Policy*, OECD, Paris.

OECD (1999), *Handbook of Incentive Measures for Biodiversity: Design and Implementation* OECD, Paris.

OECD (2002), *Handbook of Biodiversity Valuation: A Guide for Policy Makers*, OECD, Paris.

OECD (2003), *Harnessing Markets for Biodiversity Towards Conservation and Sustainable Use*, OECD, Paris.

OECD (2004), *OECD Environmental Performance Reviews: Sweden*, OECD, Paris.

OECD (2006), *OECD Environmental Performance Reviews: Korea*, OECD, Paris.

Ohl, C. *et al.* (2006), "Managing Land Use and Land Cover Change in the Biodiversity Context with Regard to Efficiency, Equality and Ecological Effectiveness", *UFZ-Discussion Papers* 3/2006, February 2006, UFZ Centre for Environmental Research Leipzig, Germany.

Okun, A.M. (1975), *Equality and Efficiency: The Big Tradeoff*, The Brookings Institution, Washington DC.

Ostrom, E. and R. Gardner (1993), "Coping with Asymmetries in the Commons: Self-Governing Irrigation Systems Can Work", *Journal of Economic Perspectives*, 7(4), 93-112.

Pagiola, S., A. Arcenas and G. Platais (2005), "Can Payments for Environmental Services Help Reduce Poverty? An Exploration of the Issues and the Evidence to Date from Latin America", *World Development* 33(2), 237-253.

Pearce, D. (1983), *Cost-Benefit Analysis*, Second edition, MacMillan, London.

Pearce, D. (1998), "Cost-benefit Analysis and Environmental Policy", *Oxford Review of Economic Policy*, 144, 84-100.

Pearce, D. (2006), "Framework for Assessing the Distribution of Environmental Quality", in Serret, Y. and N. Johnstone (eds.), *The Distributional Effects of Environmental Policy*, Edward Elgar, Cheltenham, UK.

Pearce, D. and D. Moran (1994), *The Economic Value of Biodiversity*, IUCN and Earthscan, London.

Pearce, D. and R.K. Turner (1990), *Economics of Natural Resources and the Environment*, Johns Hopkins Press, Baltimore.

Pearce, D. and D. Ulph (1995), "A Social Discount Rate For The United Kingdom", *CSERGE Working Paper* No. 95-01, School of Environmental Studies University of East Anglia, Norwich, UK.

Pearce, D., G. Atkinson and S. Mourato (2006), *Cost Benefit Analysis and the Environment: Recent Developments*, OECD, Paris.

Pearce, D. *et al.* (2003), "Valuing the Future – Recent Advances in Social Discounting", *World Economics* 4(2), 121-141.

Pelletier, M. (2002), *Enhancing Cree Participation by Improving The Forest Management Planning Process*, a project of the Waswanipi Cree Model Forest, Natural Resources Canada, Canadian Forest Service, Ottawa.

Peluso, NL. (1993), "Coercing Conservation: The Politics of State Resource Control", *Glob. Environ. Change* 3(2): 199-218.

Perrings, C. *et al.* (eds.) (1995), *Biodiversity Loss: Economic and Ecological Issues*, Cambridge University Press, Cambridge.

Pezzey, J. (1992), *Sustainable Development Concepts: An Economic Analysis*, World Bank, Washington, DC.

Pretty, J. (2003), "Social Capital and the Collective Management of Resources", *Science* 302 (12 Dec. 2003), 1912-1914.

Proctor, W. (2000), "Towards Sustainable Forest Management, An Application of Multi-criteria Analysis to Australian Forest Policy", paper presented at the *Third International Conference of the European Society for Ecological Economics*, 3-6 May 2000, Vienna, Austria.

Proctor, W. and M. Drechsler (2003), "Deliberative Multicriteria Evaluation: A case study of recreation and tourism options in Victoria Australia", paper presented at the *European Society for Ecological Economics, Frontiers 2 Conference*, Tenerife, 11-15 February 2003.

Quang, D.V. and T.N. Anh (2007), "Commercial Collection of NTFPs and Households Living in or Near the Forests: Case study in Que, Con Cuong and Ma, Tuong Duong, Nghe An, Viet Nam", *Ecological Economics*, forthcoming.

Radner, R. and J. Stiglitz (1984), "A Nonconcavity in the Value of Information," in M. Boyer and R. Kihlstrom (eds.) *Bayesian Models in Economic Theory*, Elsevier Science Publishers, New York.

Ramsey, F.P. (1928), "A Mathematical Theory of Saving", *Economic Journal* 38, 543-559.

Rangarajan, M. (1996), *Fencing the Forest: Conservation and Ecological Change in India's Central Provinces 1860-1914*, Oxford University Press, New Delhi.

Rao, M., A. Rabinowitz and S.T. Khaing (2002), "Status Review of the Protected-Area System in Myanmar, with Recommendations for Conservation Planning", *Conserv. Biol.* 16(2): 360-68.

Reardon, T. and S.A. Vosti (1995), "Links Between Rural Poverty and the Environment in Developing Countries: Asset Categories and Investment Poverty", *World Development* 23(9), 1495-1506.

Reddy, S.R.C. and S. P. Chakravarty (1999), "Forest Dependence and Income Distribution in a Subsistence Economy: Evidence from India" *World Development* 27(7), 1141-1149.

Reid, H. *et al.* (2004), "Co-management of Contractual National Parks in South Africa: Lessons from Australia", *Conservation and Society*, 2, 2: 377-409.

Reiling, S.D., H. Cheng and C. Trott (1992), "Measuring the Discriminatory Impact Associated with Higher Recreational Fees", *Leisure Science* 14(1992): 121-137.

River Dialogue (2003), *River Dialogue Newsletter* 1, September 2003, *www.riverdialogue.org.*

River Dialogue (2004), *River Dialogue Newsletter* 2, April 2004, *www.riverdialogue.org.*

Roberts, E.H. and M.K. Gautam (2003), *Community Forestry Lessons from Australia: A Review of International Case Studies*, research report presented to Faculties Research Grant Scheme 2002-2003, The Australian National University, School Resources, Environment and Society, Canberra, Australia.

 PEOPLE AND BIODIVERSITY POLICIES – ISBN 978-92-64-03431-0 – © OECD 2008

Russell, C. and W. Vaughan (1982), "The National Recreational Fishing Benefits of Water Pollution Control", *Journal of Environmental Economics and Management*, 1982, 328-354.

Saberwal, V., M. Rangarajan and A. Kothari (eds.) (2000), *People, Parks and Wildlife: Towards Co-Existence*, Orient Longman Limited, Hyderabad, India.

Sachs, J.D. and A.M. Warner (1997), "Fundamental Sources of Long-Run Growth", *American Economic Review*, 87(2), 184-88.

Schläpfer, F. and N. Hanley (2003), "Do Local Landscape Patterns Affect the Demand for Landscape Amenities Protection?" *Journal of Agricultural Economics* 54(1), 21-35.

Schläpfer, F., A. Roschewitz and N. Hanley (2004), "Validation of Stated Preferences for Public Goods: A Comparison of Contingent Valuation Survey Response and Voting Behaviour", *Ecological Economics*, 51(1/2), 1-16.

Schmidt-Soltau, K. (2003), "Conservation-related Resettlement in Central Africa: Environmental and Social Risks", *Dev. Change* 34: 525-51.

Schneider, F. (2005), "Shadow Economies of 145 Countries All over the World: What Do We Really Know?" *Crema Research Working Paper* 2005-13. Center for Research in Economics, Management and the Arts, Basel.

Schou, J.S. and J.C. Streibig (1999), "Pesticide Taxes in Scandinavia", *Pesticide Outlook* 10, Dec. 1999, 127-129.

Sen, A.K. (1997), *Choice, Welfare and Measurement*, Harvard University Press, Cambridge, MA.

Serret, Y. and N. Johnstone, (2006), *The Distributional Effects of Environmental Policy*, Edward Elgar, Cheltenham, UK.

Shyamsundar, P. and R. Kramer (1997), "Biodiversity Conservation – At What Cost? A Study of Households in the Vicinity of Madagascar's Mantadia National Park", *Ambio*, 26(3), 180-184.

Simpson, R.D., R.A. Sedjo and J.W. Reid (1996), "Valuing Biodiversity for Use in Pharmaceutical Research", *Journal of Political Economy* 104(1), 163-185.

Smith, R.J. *et al.* (2003), "Governance and the Loss of Biodiversity", *Nature* 426(6962), 67-70.

Smith, S. (1995), "*'Green' Taxes and Charges: Policy and Practice in Britain and Germany*", The Institute of Fiscal Studies, London.

Smyth, D. (2001), "Joint Management of National Parks in Australia", in Baker, R., Davies, J. and Young, E. (eds.), *Working on Country, Contemporary Indigenous Management of Australia's Lands and Coastal Regions*, Oxford University Press, Oxford, United Kingdom.

Solow, R.M. (1974), "The Economics of Resources or the Resources of Economics", *American Economic Review* 64(2), 869-877.

Southgate, D. (1998), *Tropical Forest Conservation: An Economic Assessment of the Alternatives in Latin America*, Oxford University Press, Oxford.

Southgate, D. *et al.* (2000), "Markets, Institutions and Forestry: The Consequences of Timber Trade Liberalization in Ecuador", *World Development* 28(11), 2005-2012.

Spence M. (1999), *Dispossessing the Wilderness: Indian Removal and the Making of the National Parks*, Oxford Univ. Press, New York.

Start, D. and I. Hovland (2004), *Tools for Policy Impact, A Handbook for Researchers*, Research and Policy Development Programme, Overseas Development Institute, London.

Stern, N. (1997), *Macroeconomic Policy and the Role of the State in a Changing World; Development Strategy and Management of the Market Economy*. Volume 1, Oxford University Press, Clarendon Press for the United Nations, Oxford and New York.

Stern, N. (2006), *Stern Review on the Economics of Climate Change*, HMS Treasury, London

Stoll-Kleemann, S. (2001), "Reconciling Opposition to Protected Areas Management in Europe: The German Experience", *Environment* 43(5), 32-44.

Suman, D., M. Shivlani and J.W. Milon (1999), "Perceptions and Attitudes Regarding Marine Reserves: A Comparison of Stakeholder Groups in the Florida Keys National Marine Sanctuary", *Ocean and Coastal Management*, 42: 1019-1040.

Sunderlin, W.D. *et al.* (2005), "Livelihoods, Forests, and Conservation in Developing Countries: An Overview", *World Development* 33, 9, 1383-1402.

Swanson, T. (1994), "The Economics of Extinction Revisited and Revised: A Generalized Framework for the Analysis of the Problem of Endangered Species and Biodiversity Losses", *Oxford Economic Papers* 46, 800-821.

Swanson, T. (ed.) (1995), *The Economics and Ecology of Biodiversity Decline*, Cambridge University Press, Cambridge.

Swanson, T. (1996), "The Reliance of Northern Economies on Southern Biodiversity: Biodiversity as Information", *Ecological Economics* 17(1), 1-8.

Taylor, D.F. (2001), "Employment-based Analysis: An Alternative Methodology for Project Evaluation in Developing Regions, with an Application to Agriculture in Yucatán", *Ecological Economics*, 36: 249-262.

Taylor, D.F. and I. Adelman (1996), *Village Economies: The Design, Estimation and Use of Village-wide Economic Models*, Cambridge University Press, Cambridge.

The Economist (2006), "Shots Across the Stern", *The Economist*, Economics Focus, 13 Dec. 2006.

The Economist (2007), "Conservation in Colorado", *The Economist*, 1 Feb. 2007.

Theil, H. and R. Finke (1983), "The Consumer's Demand for Diversity", *European Economic Review*, 23(3), 395-400.

Tikka, P.M. (2003), "Conservation Contracts in Habitat Protection in Southern Finland", *Environmental Science and Policy*, 6, 271-278.

Torell, D.J. (1993), "Viewpoint: Alternative Dispute Resolution in Public Management", *Journal of Range Management* 46 (6), November, 70-73.

Trannoy, A. (2003), "About the Right Weight of the Social Welfare Function when Needs Differ", *IDEP Working Papers* 2004 0304, Institut d'economie publique (IDEP), Marseille, France.

US Department of Interior, US Fish and Wildlife Service and Environmental Defense (2005a), *Conservation Profiles: Landowners Help Imperiled Wildlife*, US Fish and Wildlife Service, Washington DC.

US Department of Interior, US Fish and Wildlife Service, National Association of Conservation Districts, USDA, American Forest Foundation and Environmental Defence (2005b), *Working Together: Tolls for Helping Imperiled Wildlife on Private Lands*, US Fish and Wildlife Service, Washington DC.

PEOPLE AND BIODIVERSITY POLICIES – ISBN 978-92-64-03431-0 – © OECD 2008

UNDP (United Nations Development Programme) (1990), *Human Development Report 1990*, United Nations Development Programme, United Nations, New York.

Unsworth, R. *et al.* (2005), *Mexican Wolf Blue Range Reintroduction Project 5-Year Review, Socio-economic Component*, US Fish and Wildlife Service, Arlington, Virginia.

Warr, P.G. (1983), "The Private Provision of a Public Good is Independent of the Distribution of Income", *Economics Letters* 13, 207-211.

Wätzold, F. and M. Drechsler (2005), "Spatially Uniform *versus* Spatially Heterogeneous Compensation Payments for Biodiversity-Enhancing Land-Use Measures", *Environmental and Resource Economics* 31, 73-93.

Weimer, D.L. and A.R. Vining (1998), *Policy Analysis – Concepts and Practice*, third edition, Prentice Hall.

Weitzman, M.L. (1998), "Why the Far Distant Future Should be Discounted at its Lowest Possible Rate", *Journal of Environmental Economics and Management* 36, 201-208.

Wells, M. (1992), "Biodiversity Conservation, Affluence and Poverty: Mismatched Costs and Benefits and Efforts to Remedy Them", *Ambio* 21(3), 237-243.

Wells, M., K. Brandon and L. Hannah (1992), *People and Parks: Linking Protected Area Management with Local Communities*, The World Bank, Washington DC.

Wick, K. and E.H. Bulte (2006), "Contesting Resources – Rent Seeking Conflict and the Natural Resource Curse", *Public Choice* 128: 457-476.

Wickham, T. (1997) "Community-based Participation in Wetland Conservation: Activities and Challenges of the Danau Sentarium Wildlife Reserve Conservation Project, Danau Sentarium Wildlife Reserve, West Kalimantan, Indonesia", case study 5, in Claridge, G. and O'Callaghan (eds.), *Community Involvement in Wetland Management: Lessons from the Field*, Proceedings of Workshop 3. Wetlands, Local People and Development, International Conference on Wetlands Development, 9-13 October 1995, Kuala Lumpur, Malaysia, Wetlands International, Kuala Lumpur.

Willig, R.D. (1976), "Consumer's Surplus without Apology", *American Economic Review* 66(4), 589-97.

Wilson, R.K. (2003), "Community-Based Management and National Forests in the Western United States- Five Challenges", *Policy Matters* 12: 216-224.

World Bank (2002), *Operational Policy 4.12: Involuntary Resettlement*, The World Bank, Washington, DC.

World Bank (2006), *Strengthening Forest Law Enforcement and Governance: Strengthening a Systemic Constraint to Sustainable Development*, report No. 36638-GLB, The World Bank, Washington, DC.

Young, Z, Makoni, G and Boehmer Christiansen, S. (2001), "Green Aid in India and Zimbabwe – Conserving Whose Community?" *Geoforum* 32, 299-318.

Zbinden, S. and D.R. Lee (2005) "Paying for Environmental Services: An Analysis of Participation in Costa Rica's PSA Program", *World Development* 33(2), 255-272.

ISBN 978-92-64-03431-0
People and Biodiversity Policies
Impacts, Issues and Strategies for Policy Action

ANNEX A

Case Study Overview

Case study	Developed or developing country?	Issue	Conservation context	Key issues	Location in book (section or box)
Lucas *vs.* South Carolina Coastal Council	Developed (USA)	Conflict (court case)	Coastal beach/dune system	Importance of institutional safety valves for distributive issues	1.1.1
Natura 2000	Developed (Germany, Finland)	Conflict	Various protected areas	Delay and failure of conservation programmes due to inefficient communication	1.1.2, 5.3.5 and 6.4.2
Extractive reserves	Developing (Brazil)	Instrument choice	Forests	No income gain for locals	1.1.3
Privatisation of mangroves (conversion to agriculture and aquaculture)	Developing (Viet Nam)	Measurement of distributive impacts/Gini coefficient	Forests	Use of Gini coefficient to measure the effect of privatisation and land conversion on inequality	2.3.1
Different scenarios for the Ream National Park	Developing (Cambodia)	Comparison of different conservation scenarios/Extended cost-benefit analysis (CBA)	National park	Distributional impact on different stakeholders	2.3.2
Alternative forest management in the Upper Great Lake Regions	Developed (USA)	Measurement of distributional impacts/social accounting matrix (SAM)	Forests	Regressive policy	2.3.3
Farming and ranching in Yucatán	Developing (Mexico)	Employment based analysis and CBA	Farming	EBA shows impact of different land use scenarios on local employment. The results are different from those of CBA	2.4.1
Farming or fishing in Bangladesh	Developing (Bangladesh)	Analysis of different land-use schemes.	Floodplain	Spillover impacts of agriculture management to biodiversity and livelihood of poor	2.5.1

Case study	Developed or developing country?	Issue	Conservation context	Key issues	Location in book (section or box)
Different scenarios for forest policy in New South Wales	Developed (Australia)	Participatory multidimensional measures/multi-criteria analysis (MCA)	Forests	MCA reveals trade-offs between social, economic and biodiversity criteria in different scenarios	2.5.2
Development scenarios for the Bucco-Reef Marine Park	Developing (Tobago)	Participatory multidimensional measures/MCA with stakeholder involvement	Marine park	MCA helps compare different development scenarios using economic, social and biodiversity criteria. Stakeholders weight the criteria	2.5.2
Development scenarios in the nature reserve Šúr wetland area	Developed (Slovakia)	Participatory multidimensional measures/MCA	Wetland, nature reserve	MCA helps compare different development scenarios ranging from strict nature protection to unlimited economic development	2.5.2
Stakeholder analysis in the Royal Bardia National Park, Nepal	Developing (Nepal)	Social impact assessment with stakeholder analysis	National park	Stakeholder analysis reveals the interests, the powers, the scale of influence and the means of different groups related to the management of the park	2.5.3
Conservation easements	Developed (USA)	Instruments		Regressive	3.2.2
ÖPUL	Developed (Austria)	Instruments	Farming	Regressive	Box 3.2
Contracted conservation	Developed (Germany/EU)	Instruments	Farming		Box 5.2
Ely's citizens' jury	Developed (UK)	Participatory measures/citizens' jury	Wetland	Analysing options for wetland management with public participation using citizens' jury	6.3.2
National park management in New South Wales	Developed (Australia)	Participatory measures/citizens' jury	National park	Citizens' jury reveals citizens' opinions about the management options for a national park	6.3.2
River dialogue	Developed (Sweden, Netherlands, Estonia)	Participatory measures/focus groups	Water	EU research programme revealing opinions of stakeholders on the Water Framework Directive in three countries	6.3.2
Kalloni Bay	Developed (Greece)	Participatory measures/focus groups	Wetland	Focus group interviews show the opinions of different stakeholders on the current status of a wetland area and its further development	6.3.2
Boreal Forest Program	Developed (Canada)	Participatory measures/ scenario workshop, roundtable	Forests	National roundtable with the involvement of stakeholders examines how to balance conservation and economic activity of boreal forests	6.3.2

Case study	Developed or developing country?	Issue	Conservation context	Key issues	Location in book (section or box)
National workshop project related to the designation of critical habitats for species	Developed (USA)	Participatory measures/scenario workshops	Wildlife	Stakeholder workshops to reveal opinions on the designation of critical habitats	6.3.2
Ria Lagartos Biosphere Reserve – conflicts around restrictions	Developing (Mexico)	Conflict resolution, participatory measures	Forests	User conflict because zoning and restrictions are introduced Learning process of community involvement	6.4.2
Rights of Saami people (indigenous group)	Developed (Sweden)	Conflict, community based management/lack of participatory measures, in conflict resolution	Wildlife	Conflict around the hunting and fishing rights of indigenous people, court case	6.4.2 and 8.3.1
Wood Buffalo National Park: conflict over diseased bison	Developed (Canada)	Conflict, participatory measures/joint management	Wildlife	Conflict between park managers and native groups on managing a bison disease Successful consultation with the native group	6.4.2
Mexican Wolf Blue Range Reintroduction Project	Developed (United Sates)	Conflict, participatory measures/joint management	Wildlife	Negative impacts of the reintroduction of wolf on cattle ranchers and native tribes Conflicts resolved through consultation, court cases and administrative decisions of the stakeholder management committee	6.4.2
Compensation on Natura 2000	Developed EU countries)	Instruments/ compensation scheme	Natura 2000 sites	Mitigation of the distributive effects imposed by the EU regulation on Natura 2000 site through compensation	7.2.1
Neusiedler See-Seewinkel National Park	Developed (Austria)	Instruments/ compensation scheme	Park	Prevention of negative distributive impacts of the designation of protected area through a compensation measure	7.2.1
Safe Harbor Programme	Developed (USA)	Instruments/ voluntary schemes	Wildlife	Voluntary programme that encourages private land owners to restore and maintain habitats for endangered species, thereby avoiding further regulation	7.2.2
Habitat Stewardship Programme	Developed (Canada)	Instruments/ voluntary schemes	Wildlife	Voluntary conservation on private land with financial support	7.2.2

Case study	Developed or developing country?	Issue	Conservation context	Key issues	Location in book (section or box)
Greencover program	Developed (Canada)	Instruments/ voluntary schemes	Farming	Agri-environmental measure of Canada with biodiversity component Provision of technical and financial support	7.2.2
Agri-environmental measures in the EU	Developed (EU countries)	Instruments/ voluntary schemes	Farming	Voluntary agri-environmental programme of EU with financial assistance	7.2.2
Natural Forest Reserve Programme	Developed (Austria)	Instruments/ voluntary schemes	Forests	Voluntary programme for private forest owners on biodiversity friendly management, including a yearly payment	7.2.2
Entry Level Stewardship Scheme	Developed (UK)	Instruments/ voluntary schemes	Farming	Change in the agri-environmental scheme to attract more farmers in conservation of biodiversity, landscape, access and historic environment Financial support is provided	7.2.2
Forest Biodiversity Programme	Developed (Finland)	Instruments/ voluntary schemes	Forests	Voluntary conservation programmes for private forest owners with financial incentives	7.2.2
BushTender Program	Developed (Australia)	Instruments/ auctions		Competitive tendering process among private land owners for biodiversity management	7.2.2
Conservation Banking	Developed (USA)	Instruments/ market based mechanism	Conservation	Voluntary conservation programme for mitigating loss of protected habitats elsewhere	7.2.2
Ecological Gifts Programme	Developed (Canada)	Instruments/ donations	Conservation	Voluntary conservation activity is motivated through a tax reduction programme	7.2.2
Danau Sentarium Wildlife Reserve	Developing (Indonesia)	Participatory measures/ community based management	Wildlife	Traditional restrictions on fishing are used in the community based management	8.3.1
CAMPFIRE	Developing (Zimbabwe)	Participatory measures/ community based management	Wildlife	Giving user rights back to local communities and fostering biodiversity friendly management with financial incentives (*e.g.* benefit sharing)	8.3.1
Kakadu National Park	Developed (Australia)	Participatory measures/joint management of community and authority	Wildlife	Contract between Aboriginal people and park administration on the park management: sharing tasks and benefits	8.3.2

Case study	Developed or developing country?	Issue	Conservation context	Key issues	Location in book (section or box)
Canada's Model Forest Program	Developed (Canada)	Participatory measures/ stakeholder management	Forests	Experimental programme for sustainable forestry through the partnership of different stakeholders	8.3.3
Waswanipi Cree Model Forest	Developed (Canada)	Participatory measures/ stakeholder management	Forests	Stakeholder management for sustainable forestry on the land of a native tribe	8.3.3
National Community Forest Partnership	Developed (UK)	Participatory measures/ stakeholder management	Forests	Voluntary forest conservation programme for local forests around towns, operated in a partnership with various stakeholders.	8.3.3
Watershed Management	Developed (USA)	Participatory measures/ stakeholder management	Watersheds	Voluntary programme for the conservation for watersheds through the partnership of local stakeholders	8.3.3
Management of the Djoudj National Park	Developing (Senegal)	Participatory measures/ stakeholder management	Park	Introduction of a new participatory management of a national park with more involvement of locals in different stakeholder committees	8.3.3
Daepho River improvement	Developed (Korea)	River water quality	Marine	Local residents collaborating to restore water quality and avoid more general restrictions related to water use	8.3.3
Channeling park fees to local communities	Developing (Uganda)	Instruments/ channeling of entrance fees to local communities	National park	Benefit sharing programme: Part of the revenues from the park entrance fees are given to local communities adjacent to the park	8.3.4
Sharing buffer zone fees in Langtang National Park	Developing (Nepal)	Instruments/ channeling of entrance fees to local communities	National park	Benefit sharing programme: channelling part of the revenues from buffer zone fees to local	8.3.4

OECD PUBLICATIONS, 2, rue André-Pascal, 75775 PARIS CEDEX 16
PRINTED IN FRANCE
(97 2008 03 1 P) ISBN 978-92-64-03431-0 – No. 56121 2008